An Introduction to the Theory of Stellar Structure and Evolution

The theory of stellar structure and evolution is elegant and impressively powerful. Bringing together all branches of physics, it is able to predict how the complex internal structure of stars changes from their birth to their death, what nuclear fuel stars burn, and what their ultimate fate is – a fading white dwarf, or a cataclysmic explosion as a supernova, leaving behind a collapsed neutron star or a black hole. This lucid textbook provides students with a clear and pedagogical introduction to stellar structure and evolution. It requires only basic physics and mathematics learned in first- and second-year undergraduate studies, and it assumes no prior knowledge of astronomy.

Beginning with what is known about stars from observations, the theory of stellar evolution is then laid out mathematically and the basic physics related to the structure of stars is reviewed. Next, nucleosynthesis, simple stellar models, and the principles of stability are introduced, leading to a schematic picture of stellar evolution as a whole. This model is then developed further, bringing in what scientists have learned from computer simulations, and comparing the results with more detailed observations of main-sequence stars like the Sun, red giants, planetary nebulae, and white dwarfs. The exotic and often spectacular final stages of many stars – supernovae, pulsars, and black holes – are treated next. Finally, a global picture of the stellar evolutionary cycle is presented.

The unique feature of this book is the emphasis it places, throughout, on the basic physical principles governing stellar evolution. The processes are always explained in the simplest terms, while maintaining full mathematical rigor. Exercises and their full solutions are also included to help students test their understanding. This textbook provides a stimulating introduction for undergraduates in astronomy, physics, planetary science, and applied mathematics who are taking a course on the physics of stars.

Dina Prialnik is a Professor of Planetary Physics at Tel Aviv University. She has conducted research on binary stars at the Space Telescope Science Institute in Baltimore, Maryland, and is a member of the Working Group on Cometary Physics at the International Space Science Institute in Bern, Switzerland. She has published over eighty articles on stellar evolution and on the physics of comets. She enjoys a high reputation as a teacher of astrophysics and planetary physics to undergraduate and graduate students, and has contributed to training high-school teachers in the teaching of astronomy and to writing the Astronomy course for The Open University of Israel.

An Introduction to the Theory of Stellar Structure and Evolution

Dina Prialnik
Tel Aviv University

CAMBRIDGE UNIVERSITY PRESS
Cambridge, New York, Melbourne, Madrid, Cape Town, Singapore, São Paulo, Delhi

Cambridge University Press
32 Avenue of the Americas, New York, NY 10013-2473, USA

www.cambridge.org
Information on this title: www.cambridge.org/9780521659376

© Cambridge University Press 2000

This publication is in copyright. Subject to statutory exception
and to the provisions of relevant collective licensing agreements,
no reproduction of any part may take place without
the written permission of Cambridge University Press.

First published 2000
Reprinted 2004, 2005, 2006, 2007, 2008

Printed in the United States of America

A catalog record for this publication is available from the British Library.

Library of Congress Cataloging in Publication Data
Prialnik, Dina.
An introduction to the theory of stellar structure and evolution / Dina Prialnik.
p. cm.
Includes bibliographical references.
ISBN 0-521-65065-8 (hb) – ISBN 0-521-65937-X (pb)
1. Stars – Structure. 2. Stars – Evolution. I. Title.
QB808 .P75   2000
523.8 21 – dc21   99-044948

ISBN 978-0-521-65065-6 hardback
ISBN 978-0-521-65937-6 paperback

Cambridge University Press has no responsibility for
the persistence or accuracy of URLs for external or
third-party Internet Web sites referred to in this publication
and does not guarantee that any content on such
Web sites is, or will remain, accurate or appropriate.

To my son

Contents

Preface	*page* xi
1 Observational background and basic assumptions	**1**
1.1 What is a star?	1
1.2 What can we learn from observations?	2
1.3 Basic assumptions	6
1.4 The H–R diagram: a tool for testing stellar evolution	9
2 The equations of stellar evolution	**15**
2.1 Local thermodynamic equilibrium	16
2.2 The energy equation	17
2.3 The equation of motion	19
2.4 The virial theorem	21
2.5 The total energy of a star	24
2.6 The equations governing composition changes	26
2.7 The set of evolution equations	29
2.8 The characteristic timescales of stellar evolution	30
3 Elementary physics of gas and radiation in stellar interiors	**35**
3.1 The equation of state	36
3.2 The ion pressure	38
3.3 The electron pressure	39
3.4 The radiation pressure	43
3.5 The internal energy of gas and radiation	44
3.6 The adiabatic exponent	45
3.7 Radiative transfer	48

4 Nuclear processes that take place in stars — 53
- 4.1 The binding energy of the atomic nucleus — 53
- 4.2 Nuclear reaction rates — 56
- 4.3 Hydrogen burning I: the p–p chain — 59
- 4.4 Hydrogen burning II: the CNO bi-cycle — 62
- 4.5 Helium burning: the triple-α reaction — 63
- 4.6 Carbon and oxygen burning — 65
- 4.7 Silicon burning: nuclear statistical equilibrium — 67
- 4.8 Creation of heavy elements: the s- and r-processes — 68
- 4.9 Pair production — 70
- 4.10 Iron photodisintegration — 70

5 Equilibrium stellar configurations – simple models — 72
- 5.1 The stellar structure equations — 72
- 5.2 What is a simple stellar model? — 73
- 5.3 Polytropic models — 74
- 5.4 The Chandrasekhar mass — 79
- 5.5 The Eddington luminosity — 81
- 5.6 The standard model — 82
- 5.7 The point-source model — 86

6 The stability of stars — 90
- 6.1 Secular thermal stability — 90
- 6.2 Cases of thermal instability — 92
- 6.3 Dynamical stability — 95
- 6.4 Cases of dynamical instability — 97
- 6.5 Convection — 98
- 6.6 Cases of convective instability — 101
- 6.7 Conclusion — 104

7 The evolution of stars – a schematic picture — 106
- 7.1 Characterization of the ($\log T$, $\log \rho$) plane — 107
- 7.2 The evolutionary path of the central point of a star in the ($\log T$, $\log \rho$) plane — 111
- 7.3 The evolution of a star, as viewed from its centre — 115
- 7.4 The theory of the main sequence — 118

	7.5	Outline of the structure of stars in late evolutionary stages	124
	7.6	Shortcomings of the simple stellar evolution picture	128
8	**The evolution of stars – a detailed picture**		**131**
	8.1	The Hayashi zone and the pre-main-sequence phase	132
	8.2	The main-sequence phase	138
	8.3	Solar neutrinos	143
	8.4	The red giant phase	146
	8.5	Helium burning in the core	151
	8.6	Thermal pulses and the asymptotic giant branch	155
	8.7	The superwind and the planetary nebula phase	160
	8.8	White dwarfs – the final state of nonmassive stars	164
	8.9	The evolution of massive stars	170
	8.10	The H–R diagram: Epilogue	173
9	**Exotic stars: supernovae, pulsars, and black holes**		**176**
	9.1	What is a supernova?	176
	9.2	Supernova explosions – the fate of massive stars	180
	9.3	Nucleosynthesis during supernova explosions	184
	9.4	Supernova progenies: neutron stars – pulsars	187
	9.5	Very massive stars and black holes	191
	9.6	The luminosity of accretion and hard radiation sources	192
10	**The stellar life cycle**		**195**
	10.1	The interstellar medium	195
	10.2	Star formation	196
	10.3	Stars, brown dwarfs, and planets	199
	10.4	The initial mass function	203
	10.5	The global stellar evolution cycle	208

Appendices

   1 The equation of radiative transfer   215
   2 Solutions to all the exercises   223
   3 Physical and astronomical constants   249

Bibliography   251

Index   255

Preface

For over ten years I have been teaching an introductory course in astrophysics for undergraduate students in their second or third year of physics or planetary sciences studies. In each of these classes, I have witnessed the growing interest and enthusiasm building up from the beginning of the course toward its end.

It is not surprising that astrophysics is considered interesting; the field is continually gaining in popularity and acclaim due to the development of very sophisticated telescopes and to the frequent space missions, which seem to bring the universe closer and make it more accessible. But students of physics have an additional reason of their own for this interest. The first years of undergraduate studies create the impression that physics is made up of several distinct disciplines, which appear to have little in common: mechanics, electromagnetism, thermodynamics, and atomic physics, each dealing with a separate class of phenomena. Astrophysics – in its narrowest sense, as *the physics of stars* – presents a unique opportunity for teachers to demonstrate and for students to discover that complex structures and processes do occur in Nature, for the understanding of which all the different branches of physics must be invoked and combined. Therefore, a course devoted to the physics of stars should perhaps be compulsory, rather than elective, during the second or third year of physics undergraduate studies. The present book may serve as a guide or textbook for such a course.

Books on astrophysics fall mostly into two categories: on the one hand, extensive introductions to the field covering all its branches, from planets to galaxies and cosmology, quite often including an introduction to the main fields of physics as well; and on the other hand, specialized books, often including up to date results of ongoing studies. The former are aimed at readers who have not yet received any real training in physics; the latter, at graduate students who are specializing in astronomy or astrophysics. The present book is aimed at students who fall between these extremes: undergraduates who have acquired a basic mathematical background and have been introduced to the basic laws of physics during the first two or three semesters of studies, but have no prior knowledge of astronomy.

The purpose of this book is to satisfy the eagerness to comprehend the realm of stars, by focusing on fundamental principles. The students are made to understand, rather than become familiar with, the different types of stars and their evolutionary trends. As far as possible, I have refrained from burdening the reader with astronomical concepts and details, in an attempt to make the text suitable for students of physics who do not necessarily intend to pursue astrophysics any further. Thus, odd as it may seem, there is no mention of concepts that are so familiar to astronomers, such as magnitude, colour index, spectral class, and so forth. Equally odd may appear the use of SI units, which is still alien to astrophysics, but has become common, in fact mandatory, in physics studies. I have complied with this demand, despite my conviction that, perhaps surprisingly, astrophysicists still think in terms of cgs units. (One hardly comes across stellar opacities expressed in square metres per kilogram, or densities in kilograms per metre cubed.) As is customary in textbooks, exercises are scattered throughout the book and solutions are provided in an appendix.

The theory of stellar evolution is developed in a methodical manner. The student is led step by step from the formulation of the problem to its solution on a path that appears very natural, even obvious at times. I have tried to avoid the widely adopted alternative of following the progress of a star's evolution, enumerating the different phases with their inherent physical aspects. I find the *logical*, rather than the *chronological*, method the best way of presenting this theory, the way any other established theory is usually presented. When each chapter of a scientific book relies on the preceding one and leads to the next, there is hope of arousing in the reader sufficient curiosity for reading on. The fascinating history of the theory of stellar structure and evolution is sometimes alluded to in "Notes" and quotations.

The first chapter introduces the subject of stellar evolution, as it arises from observations: the problem is defined and the basic assumptions (axioms) are laid down. The following six chapters are essentially theoretical: the second formulates the problem mathematically by introducing the equations of stellar evolution; the third summarizes briefly the basic physical laws involved in the study of stellar structure, serving for reference later on. Chapters 4, 5, and 6 – dealing with nucleosynthesis in stars, simple stellar models, and stability – build up to Chapter 7, which is the heart of this book. Combining the material of Chapters 3–6, it presents a general, almost schematic picture of the evolution of stars in all its aspects. From my experience, this picture remains imprinted in the students' minds long after the details have faded away. Chapter 8 is, in a way, a recapitulation of the previous chapter from a different angle: the story of stellar evolution is retold, filling in many details, as it emerges from numerical computations. Emphasis is now put on comparison with observations, thereby closing the circuit opened in Chapter 1. The next chapter deals with special objects: supernovae and their remnants, pulsars, black holes (very briefly), and other radiation sources. Finally,

Chapter 10 touches on the global picture of the stellar evolution cycle, from the galactic point of view.

I have tried to give proper credit where it was due, but occasionally I may have failed or erred. I apologize for any such failure or error, my only defence being that it was not intentional. I have refrained from referring to original papers in the text, in order not to interfere with fluency. A selection of references (by no means complete) is given in the bibliography.

Enthusiasm toward a subject of study is instigated not only by the subject itself, but quite often by the teacher. In this respect I was lucky to have been introduced to astrophysics by Giora Shaviv, and I hope to have carried on some of his passion to my own students. Computing and numerical modeling, on which the subject matter of this book relies, are not merely a skill but a true art of unique beauty and elegance. For having introduced me to this art long ago and for having been a constant source of encouragement and advice during the writing of this book, I am grateful to my husband (and former teacher) Attay Kovetz. I would like to express my gratitude and appreciation to Leon Mestel for a very careful and thorough reading of the original manuscript. This book has tremendously benefited from his countless observations, comments, and suggestions. Special thanks are due to Michal Semo and her team at the Desktop Publishing unit of Tel Aviv University for their skillful and painstaking graphics work, not to mention their endless patience and cheerfulness. Above all, I am grateful to my son Ely for gracefully bearing with a busy and preoccupied mother during the rather demanding years of adolescence.

Dina Prialnik

Tel Aviv, June 1999

CHOOSE SOMETHING LIKE A STAR

O Star, (the fairest one in sight),
We grant your loftiness the right
To some obscurity of cloud —
It would not do to say of night,
Since dark is what brings out your light.
Some mystery becomes the proud.
But to be wholly taciturn
In your reserve is not allowed.
Say something to us we can learn
By heart and when alone repeat.
Say something! And it says "I burn."
But say with what degree of heat.
Talk Fahrenheit, talk Centigrade.
Use language we can comprehend.
Tell us what elements you blend.
It gives us strangely little aid,
But does tell something in the end.
. . .

Robert Frost

1

Observational background and basic assumptions

1.1 What is a star?

A *star* can be defined as a body that satisfies two conditions: (a) it is bound by self-gravity; (b) it radiates energy supplied by an internal source. From the first condition it follows that the shape of such a body must be spherical, for gravity is a spherically symmetric force field. Or, it might be spheroidal, if axisymmetric forces are also present. The source of radiation is usually nuclear energy released by fusion reactions that take place in stellar interiors, and sometimes gravitational potential energy released in contraction or collapse. By this definition, a *planet*, for example, is not a star, in spite of its stellar appearance, because it shines (mostly) by reflection of solar radiation. Nor can a *comet* be considered a star, although in early Chinese and Japanese records, comets belonged with the "guest *stars*" – those stars that appeared suddenly in the sky where none had previously been observed. Comets, like planets, shine by reflection of solar radiation and, moreover, their masses are too small for self-gravity to be of importance.

A direct implication of the definition is that stars must evolve: as they release energy produced internally, changes necessarily occur in their structure or composition, or both. This is precisely the meaning of *evolution*. From the above definition we may also infer that the *death* of a star can occur in two ways: violation of the first condition – self-gravity – meaning breakup of the star and scattering of its material into interstellar space, or violation of the second condition – internally supplied radiation of energy – that could result from exhaustion of the nuclear fuel. In the latter case, the star fades slowly away, while it gradually cools off, radiating the energy accumulated during earlier phases of evolution. Eventually, it will become extinct, disappearing from the field of view of even the most powerful telescopes. This is what we call a *dead star*. We shall see that most stars end their *lives* by a combination of these two processes: partial breakup (or shedding of matter) and extinction. As to the *birth* of a star, this is a complex process, which presents many problems that are still under intensive investigation. We shall deal with this phase only briefly, mainly by pointing out the circumstances under which it is expected to occur.

We shall therefore start pursuing the evolution of a star from the earliest time when both conditions of the definition have been fulfilled, and we shall stop when at least one condition has ceased to be satisfied, completely and irreversibly. Finally, we shall consider the life cycle of stellar populations and the effect of stellar evolution on the evolution of galaxies within which stars reside. Galaxies are large systems of stars (up to 10^{11} or so), which also contain interstellar clouds of gas and dust. Many of the stars in a galaxy are aggregated in clusters, the largest among them containing more than 10^5 stars. The object of reference in stellar physics is, naturally, the Sun, and in galactic physics, the Galaxy to which it belongs, also known as the Milky Way galaxy.

1.2 What can we learn from observations?

Astrophysics (the physics of stars) does not lend itself to experimental study, as do the other fields of physical science. We cannot devise and conduct experiments in order to test and validate theories or hypotheses. Validation of a theory is achieved by accumulating observational evidence that supports it and its predictions or inferences. The evidence is derived from events that have occurred in the past and are completely beyond our control. The task is rather similar to that of a detective. As a rule of thumb, a theory is accepted as valid (or at least highly probable) if it withstands two radically different and independent observational tests, and of course, as long as no contradictory evidence has been found.

The information we can gather from an individual star is quite restricted. The primary characteristic that can be measured is the *apparent brightness*, which is the amount of radiation from the star falling per unit time on unit area of a collector (usually, a telescope). This radiation flux, which we shall denote I_{obs}, is not, however, an intrinsic property of the observed star, for it depends on the distance of the star from the observer. The stellar property is the *luminosity L*, defined as the amount of energy radiated per unit time – the power of the stellar engine. Since L is also the amount of energy crossing, per unit time, a spherical surface area at the distance d of the observer from the star, the measured apparent brightness is

$$I_{\text{obs}} = \frac{L}{4\pi d^2}, \tag{1.1}$$

and L may be inferred from I_{obs} if d is known. The luminosity of a star is usually expressed relative to that of the Sun, the *solar luminosity* $L_\odot = 3.85 \times 10^{26}$ J s^{-1}. Stellar luminosities range between less than $10^{-5} L_\odot$ and over $10^5 L_\odot$.

Note: The only *direct* method of determining distances to stars (and other celestial bodies) is based on the old concept of *parallax* – the angle between the

1.2 What can we learn from observations?

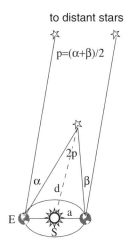

Figure 1.1 Sketch of the parallax method for measuring distances to stars.

lines of sight of a star from two different positions of the observer. The lines of sight and the line connecting the observer's positions form a triangle, with the star at the apex, as shown in Figure 1.1. The larger the distance to the object, the wider the baseline required for obtaining a discernible parallax: for objects within the solar system, distant points on Earth suffice; for stars, a much larger baseline is needed. This is provided by the Earth's orbit around the Sun, yielding a maximal baseline of $\sim 3 \times 10^{11}$ m, twice the Earth–Sun distance $a (=1$ AU$)$. Thus, the stellar parallax is obtained by determining a star's position relative to very distant, fixed stars, at an interval of half a year. Even so, the triangle obtained is very nearly isosceles, with almost right base angles, while the parallax p, defined as half the apex angle, is less than $1''$ (the largest known stellar parallax is that of *Proxima Centauri* – the star closest to our Sun, $p = 0''.76$). Consequently, to a good approximation, $d \approx a/p$. Based on this method, distances of up to about 500 light-years may be directly measured. (One light-year, 9.46×10^{15} m, is the distance travelled in one year at the speed of light.) A common astronomical unit for measuring distances, called *parsec*, is based on the parallax method: as its name indicates, it is the distance corresponding to a parallax of $1''$, amounting to about 3 light-years. Recently, the number of stars for which we have accurate distances has grown a hundredfold as a result of the activity of the satellite specially designed for this task, *Hipparcos* (high precision parallax collecting satellite), named after the greatest astronomer of antiquity, Hipparchus of Nicea (second century B.C.), who measured the celestial positions and brightnesses of almost a thousand stars and produced the first star catalogue. The satellite *Hipparcos*, which operated during 1989–1993, gathered data on more than a million nearby stars. But on the astronomical scale, distances that can be directly measured are quite small, and hence *indirect* methods have to be devised, some of which are based on the theory of stellar structure and evolution, as we shall see in Chapter 8.

The surface temperature of a star may be obtained from the general shape of its spectrum, the *continuum*, which is very similar to that of a blackbody. The *effective temperature* of a star T_{eff} is thus defined as the temperature of a blackbody that would emit the same radiation flux. It provides a good approximation to the temperature of the star's outermost layer, called the *photosphere*, where the bulk of the emitted radiation originates. If R is the stellar radius, the surface flux is $L/4\pi R^2$, and hence

$$\sigma T_{\text{eff}}^4 = \frac{L}{4\pi R^2}, \tag{1.2}$$

where σ is the Stefan–Boltzmann constant. Thus

$$L = 4\pi R^2 \sigma T_{\text{eff}}^4. \tag{1.3}$$

The surface temperatures of stars range between a few thousand to a few hundred thousand degrees Kelvin (K), the wavelength of maximum radiation λ_{max} shifting, according to Wien's law,

$$\lambda_{\text{max}} T = \text{constant}, \tag{1.4}$$

from infrared to soft X-rays. The effective temperature of the Sun is 5780 K. We should bear in mind, however, that conclusions regarding internal temperatures cannot be drawn from surface temperatures without a theory.

The chemical composition, too, can be inferred from the spectrum. Each chemical element has its characteristic set of spectral lines. These lines can be observed in the light received from stars, superimposed upon the continuous spectrum, either as emission lines, when the intensity is enhanced, or as absorption lines, when it is diminished. The elements that make up the photosphere of a star, which emits the observed radiation, may thus be identified in the stellar spectrum. But since the photosphere is very thin, the deduced composition is not representative of the bulk, opaque interior of the star. Most of the chemical elements were found to be present in the solar spectrum. In fact, the existence of the element *helium* was first suggested by spectral lines from the Sun (in the 1860s); its name is derived from "helios," the Greek word for Sun.

Under certain conditions, the mass of a star that is member of a binary system can be calculated, based on spectral line shifts. Very seldom, in eclipsing binary systems, may the radius of a star be directly derived; it can, however, be estimated from the independently derived luminosity (when possible) and effective temperature using equation (1.3). Stellar masses and radii are measured in units of the *solar mass*, $M_\odot = 1.99 \times 10^{30}$ kg, and the *solar radius*, $R_\odot = 6.96 \times 10^8$ m. The mass range is quite narrow – between $\sim 0.1 M_\odot$ and a few tens $M_\odot$; stellar radii vary typically between less than $0.01 R_\odot$ to more than $1000 R_\odot$. Much more compact stars exist, though, with radii of a few tens of kilometers.

1.2 What can we learn from observations?

Exercise 1.1: Consider a system of two stars that revolve about their centre of mass in circular orbits, for which it is possible to separate the spectral lines of the two components (known as a *spectroscopic binary system*). As a result of the Doppler effect, the lines shift periodically about a mean to shorter and longer wavelengths, as each star moves toward or away from the observer. From these shifts it is possible to determine the orbital period P_{orb} and the velocity components along the line of sight, $v_{o,1}$ and $v_{o,2}$, respectively. Denoting the angle of inclination of the orbital plane with respect to the observer by i, find the masses of the two stars, M_1 and M_2, in terms of the observables and $\sin i$.

Beside being sparse, the information one can gather is confined to a very brief moment in a star's life, even if observations are carried on for hours or years, or, hypothetically, hundreds of years. To illustrate this point, let us compare the life span of a star to that of a human being: uninterrupted observation of a star since, say, the discovery of the telescope some four hundred years ago would be tantamount to watching a person for about three minutes! Obviously, it would be impossible to learn anything (directly) about the evolution of the star from such a fleeting observation. The body of data available to the astrophysicist consists of accumulated momentary information on a very large number of stars, at different evolutionary stages. From these data, the astrophysicist is required to form a scenario describing the evolution of a single star.

Imagine, for comparison, an explorer who has never seen human beings, trying to figure out the nature and evolutionary course of these creatures, based solely on a large sample of photographs of many different humans chosen at random. The explorer will find that humans differ in many properties, such as height, colour of skin, etc., and will note, for example, that the height of the majority varies within a narrow range around a mean of, say, 1.75 m, and only the height of a small minority is significantly below this mean. These findings may be interpreted in two ways: (a) humans are intrinsically different, the tall ones being more numerous than the short ones; (b) humans are similar to one another, but their properties change in the course of their lives, their height either increasing or decreasing with age (one would not be able to tell which). In the latter case, based on the hypothesis that humans evolve, it may also be inferred that individual human beings are tall for a longer part of their lives than they are short. It might even be possible to calculate the rate of change of the human height from the relative number of individuals in different height ranges.

In a similar manner, if we find that a certain property is common to a great number of stars, we may infer – on the basis of the evolution hypothesis – that such a property prevails in stars for long periods of time. By the same token, rarely observed phenomena might not be rare events, but simply short-lived ones. At the same time, the possibility of actually rare phenomena cannot be entirely ruled out.

This is a sample of the problems one would have to face, if the understanding of stars and their evolution were to rest entirely on observation.

As the information available for any given star is so limited, the theory of stellar evolution is not meant to describe in detail the structure and expected evolutionary course of any individual star (with the exception of the Sun). Its purpose is rather to construct a general model that explains the large variety of stellar types, as well as the relations between different stellar properties revealed by observations (such as the correlation between luminosity and surface temperature, or between luminosity and mass, which we shall shortly encounter).

1.3 Basic assumptions

Guided by the observational evidence, we may add several fundamental assumptions (or axioms) to the general definition of a star, on which to base the theory of stellar structure and evolution.

Isolation

Regarding its structure and evolution, a single star may be considered isolated in empty space, although it is invariably a member of a large group – a galaxy – or even a denser group within a galaxy – a stellar cluster. (We exclude from the present discussion binary stars – a pair of stars that form a bound system.) Consequently, the initial conditions will exclusively determine the course of a star's evolution. Thus the evolutionary process of a star (metaphorically termed *life*) differs from that of live creatures, the latter being influenced to a large extent by interaction with their environment. To better grasp the isolation of stars, consider the star closest to our Sun (*Proxima Centauri*), which is at a distance of 4.3 light-years. This distance is larger than the solar diameter by a factor of 3×10^7. Such a situation would be similar to nearest neighbours on Earth being separated by a distance 3×10^7 times their height, which roughly amounts to 50 000 km. This is four times the Earth diameter or one seventh of the distance to the Moon. We *would* call this isolation! Both the gravitational field and the radiation flux, which vary in proportion to $1/d^2$, are diminished by a factor of at least $1/(3 \times 10^7)^2 \sim 10^{-15}$ from one star to another.

Uniform initial composition

A star is born with a given mass and a given, presumably homogeneous, composition. The latter depends on the time of formation and on the location within the galaxy where the star is formed. The composition of stars has been a question of intense debate for a long time. It turned out, finally, that most of the material of a newly formed star, about 70% of its mass, consists of hydrogen. The second most important element is helium, amounting to 25–30% of the mass, and there are traces of heavier elements, of which the most abundant are oxygen, carbon, and nitrogen (in that order), known collectively as the CNO group. In the Sun, for example, for every 10 000 hydrogen atoms, there are about 1000 helium atoms, eight oxygen atoms, almost four carbon atoms, one atom of nitrogen, one of neon, and less than one atom of each of the other species. The composition of stellar material is usually described by the *mass fractions* of different elements, the mass of each element per unit mass of material. It is common to denote the mass fraction of hydrogen by X, that of helium by Y, and the total mass fraction of all the other elements by Z, so that $X + Y + Z = 1$.

> **Exercise 1.2:** Calculate the mass fractions of hydrogen, helium, carbon, oxygen, nitrogen, and neon in the Sun.

With very few exceptions, the abundances of the chemical elements, as derived from stellar spectra, are remarkably similar. Moreover, they are very similar to those prevailing in the interstellar medium. As stars are born in interstellar clouds, and the composition of their surface layers is the least affected by evolutionary processes, it may be concluded that there is little difference in the *initial* composition of stars. The largest differences occur for the abundances of the heavy elements, which vary among different stars between less than 0.001 to a few percent of the entire stellar mass. But differences in the initial abundances of these elements are of secondary importance to stellar evolution. For simplicity, we shall ignore differences in the initial composition of stars. In numerical examples we shall generally adopt the solar composition. The fate of a star will then be solely dependent upon its initial mass M.

Spherical symmetry

Departure from spherical symmetry may be caused by rotation or by the star's own magnetic field (since by the first assumption, we have excluded all possible external force fields). In the overwhelming majority of cases, the energy associated with these factors is much smaller than the gravitational binding energy. We know, for example, that the period of revolution of the Sun around its axis is about 27 days, so that its angular velocity is $\omega \simeq 2.5 \times 10^{-6}$ s^{-1}. The spin velocity of more distant stars can be deduced from the broadening of spectral lines caused by the Doppler effect. The kinetic energy of rotation relative to the gravitational binding energy is of the order

$$\frac{M\omega^2 R^2}{GM^2/R} = \frac{\omega^2 R^3}{GM} \sim 2 \times 10^{-5},$$

where G is the constant of gravitation. (This is also the ratio of the centrifugal acceleration to the gravitational acceleration at the equator.)

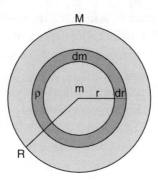

Figure 1.2 The relationship between space variables r and m in spherical symmetry.

The magnetic fields of stars similar to the Sun range from a few thousandths to a few tenths of a tesla. The larger ones may be directly deduced from split spectral lines caused by the Zeeman effect, whose separation can be measured. The energy density associated with a magnetic field B is $B^2/2\mu_0$, while the gravitational energy density is of the order of GM^2/R^4; for the Sun, even taking $B = 0.1$ T (typical of sunspots, but larger than the average magnetic field), we have

$$\frac{B^2/\mu_0}{GM^2/R^4} = \frac{B^2 R^4}{\mu_0 GM^2} \sim 10^{-11}.$$

Compact stars tend to have higher magnetic fields, but their small radii (large binding energies) compensate for them. Hence, magnetic effects on the structure of a star can usually be ignored.

Neglecting deviations from spherical symmetry, the physical properties within a star change only with the radial distance r from the centre, but they are uniform over a spherical surface of radius r. The spatial variable r may be replaced by the mass m enclosed in a sphere of radius r, as shown in Figure 1.2. The transformation between these variables is given in terms of the density ρ:

$$m(r) = \int_0^r 4\pi r^2 \rho(r) dr$$

or, in differential form,

$$dm = \rho \, 4\pi r^2 dr. \tag{1.5}$$

The advantage of using m instead of r in calculations of the changing stellar structure is that its range of variation is bounded, $0 \le m \le M$, whereas the radius may change by several orders of magnitude in the course of evolution of a star.

1.4 The H–R diagram: a tool for testing stellar evolution

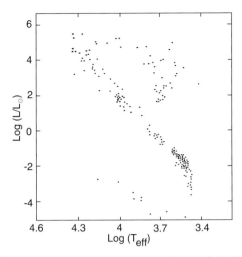

Figure 1.3 The H–R diagram of stars in the neighbourhood of the Sun.

Exercise 1.3: In a star of mass M, the density decreases from the centre to the surface as a function of radial distance r, according to

$$\rho = \rho_c \left[1 - \left(\frac{r}{R} \right)^2 \right],$$

where ρ_c is a given constant and R is the star's radius. (a) Find $m(r)$. (b) Derive the relation between M and R. (c) Show that the average density of the star (total mass divided by total volume) is $0.4\rho_c$.

1.4 The H–R diagram: a tool for testing stellar evolution

As we have seen, the two most fundamental properties of a star that can be inferred from observation are the luminosity L and the effective temperature T_{eff}. It is only natural that a possible correlation between them be sought. This was initiated independently by two astronomers at about the same time: Ejnar Hertzsprung in 1911 and Henry Norris Russell in 1913. Hence the diagram whose axes are the (decreasing) surface temperature (or related properties) and the luminosity (or related properties) bears their names, being known as *the H–R diagram*. Each observed star is represented by a point in such a diagram, an example of which is given in Figure 1.3. The results depend to some extent on the criterion used for choosing the sample of stars, for example, stars within a limited volume in the solar neighbourhood, or members of a given star cluster, or stars of apparent brightness greater than a prescribed limit, etc. The question we are interested in is whether something can be learned from this diagram regarding the evolution of stars.

It is immediately obvious from the examination of *any* H–R diagram that only certain combinations of L and T_{eff} values are possible (*a priori* there is nothing to impose such a constraint): most points are found to lie along a thin strip that runs diagonally through the (log L, log T_{eff}) plane. This strip is called *the main sequence*, and the corresponding stars are known as *main-sequence stars*.

Another populated area of the diagram is found to the right and above the main sequence: it represents stars that are brighter than main-sequence stars of same T_{eff}, or of lower T_{eff} for the same L, meaning that their spectrum is shifted toward longer wavelengths and their colour is reddish. A higher L and lower T_{eff} implies, according to equation (1.3), a large radius. Such stars are therefore called *red giants*. Their radii may attain several hundred solar radii and even more. If the Sun were to become a red giant, it would engulf the Earth and reach beyond Mars.

Another region of the (log L, log T_{eff}) plane that is relatively rich in points is located at the lower left corner: low luminosities and high effective temperatures. Stars that fall in this region have a small radius and a bluish-white colour; accordingly, they are named *white dwarfs*. White dwarf radii are of the order of the Earth's, although their masses are close to the Sun's. The typical densities of such stars are therefore tremendous; one cubic centimetre of white dwarf material would weigh more than a ton on Earth.

There are points outside these three main regions and there are conspicuously empty spots within densely populated areas of the diagram, but we shall ignore them for the moment and concentrate on the three main ones. What, if anything, can we learn from them? We recall that, in view of our basic assumptions, stars may differ from one another only in their initial mass and their age. We can therefore interpret the H–R diagram in two different ways.

1. The scatter of points is due to different ages of the stars. The implied assumption in this case is that the stars were formed at different times, and hence there are "old" stars and "young" stars. According to this hypothesis the evolution of a star can be traced in the H–R diagram by some line, with the time elapsed from the formation of the star being the changing parameter along it. Looking at a large sample of stars, each one is caught at a different age – hence the scatter of points in the diagram.
2. The properties of a star, in particular its luminosity and surface temperature, depend strongly upon its mass, the only distinguishing parameter at birth. Thus, different points in the diagram represent different stellar masses.

This is the same dilemma our earlier explorer of the human race was faced with: are the observed differences inherent or evolutionary? The explorer would have been able to choose the correct explanation, if sets of snapshots of humans of the same age, for example, pupils of different school grades, were supplied. The explorer would have immediately concluded that height is determined by age,

1.4 The H–R diagram: a tool for testing stellar evolution

Figure 1.4 Stellar clusters: *top*: the young open (amorphous shape) Pleiades cluster; *bottom*: the old globular cluster 47 Tucanae [copyright Anglo-Australian Observatory/ Royal Observatory Edinburgh; photographs by D. Malin].

whereas skin colour is an innate property. Similarly, the astrophysicist is aided by H–R diagrams of star clusters. Stars within a cluster are formed more or less simultaneously, by fragmentation of a large gas cloud (as will be explained in Section 10.2). Images of star clusters are shown in Figure 1.4. Examples of H–R diagrams of such clusters are given in Figure 1.5. We note that the main sequence

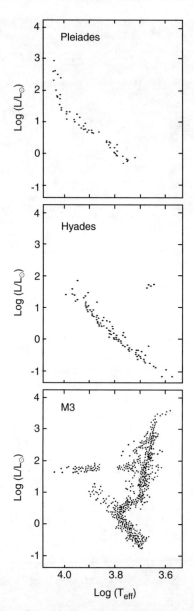

Figure 1.5 The H–R diagram of star clusters: *top:* the Pleiades cluster [adapted from H. L. Johnson & W. W. Morgan (1953), *Astrophys. J.*, 117]; *middle:* the Hyades cluster [adapted from H. L. Johnson (1952), *Astrophys. J.*, 116]; *bottom:* the globular cluster M3 [adapted from H. L. Johnson & A. R. Sandage (1956), *Astrophys. J.*, 124].

ends at different luminosities in each cluster: in one case it extends up to very high luminosities; in another case it is shorter, but at the same time there appear some red giants, which were absent in the first cluster; in yet another one, the main sequence is shorter still, and red giants are numerous. Generally, as the main sequence is depleted, the red giant and white dwarf branches are enriched. The lower part of the main sequence is always present and equally populated, allowing

1.4 The H–R diagram: a tool for testing stellar evolution

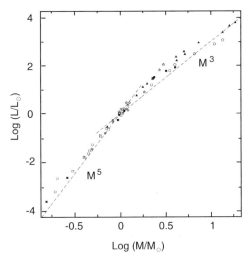

Figure 1.6 The mass–luminosity relation for main-sequence stars. Symbols denote ordinary binary stars (squares); eclipsing variables (triangles); Cepheids (pentagons); double-star statistics (stars).

for observational constraints. We may therefore conclude that being on or outside the main sequence is determined by age, whereas the location of a star along the main sequence is determined by its initial mass.

We are still unable to trace the evolutionary trajectory, whether toward or away from the main sequence, as long as the cluster ages are not determined; but the second inference can be tested. We may choose main-sequence stars with known masses and look for a correlation between their masses and luminosities. This is shown in Figure 1.6, which demonstrates that indeed there is a power-law dependence of a main-sequence star's luminosity upon its mass:

$$L \propto M^\nu,$$

with ν ranging between 3 and 5. As the Sun is a main-sequence star, the relation can be calibrated to read

$$\frac{L}{L_\odot} = \left(\frac{M}{M_\odot}\right)^\nu. \tag{1.6}$$

Do stars leave the main sequence to become red giants? Do they later turn into white dwarfs? Or, do some stars become red giants and others white dwarfs? Why are there always – in all H–R diagrams – lower main-sequence stars? Why are some changes in the stellar structure so rapid as to leave a blatant gap in the H–R diagram? Observation alone is incapable of providing answers to all these questions. We must resort to theory and use the observations that have guided us so far, in particular the H–R diagram, as a test. Here Martin Schwarzschild's words come to mind:

> "If simple perfect laws uniquely rule the universe, should not pure thought be capable of uncovering this perfect set of laws without having to lean on the crutches of tediously assembled observations? True, the laws to be discovered may be perfect, but the human brain is not. Left on its own, it is prone to stray, as many past examples sadly prove. In fact, we have missed few chances to err until new data freshly gleaned from nature set us right again for the next steps. Thus pillars rather than crutches are the observations on which we base our theories; and for the theory of stellar evolution these pillars must be there before we can get far on the right track."
>
> Martin Schwarzschild: Structure and Evolution of the Stars, 1958

Our aim throughout most of the following chapters will be to develop a theory of the stellar structure and evolution based on the laws of physics. This should ultimately lead to a theoretical H–R diagram, to be confronted with the observational one.

2

The equations of stellar evolution

We have learned a star to be a radiating gaseous sphere, made predominantly of hydrogen and helium. Radiation may be regarded as a photon gas, each "particle" carrying a quantum of energy $h\nu$, proportional to the frequency ν of the associated electromagnetic wave, and a momentum $h\nu/c$, where h is Planck's constant and c is the speed of light. This mixture of gases that makes up a star is governed by frequent collisions between its particles, ions, electrons, and photons alike. This is how Sir Arthur Eddington describes *the inside of a star*:

> "... Try to picture the tumult! Dishevelled atoms tear along at 50 miles a second with only a few tatters left of their elaborate cloaks of electrons torn from them in the scrimmage. The lost electrons are speeding a hundred times faster to find new resting-places. Look out! there is nearly a collision as an electron approaches an atomic nucleus; but putting on speed it sweeps round it in a sharp curve. A thousand narrow shaves happen to the electron in 10^{-10} of a second; sometimes there is a slide-slip at the curve, but the electron still goes on with increased or decreased energy. Then comes a worse slip than usual; the electron is fairly caught and attached to an atom, and its career of freedom is at an end. But only for an instant. Barely has the atom arranged the new scalp on its girdle when a quantum of æther waves [photon] runs into it. With a great explosion the electron is off again for further adventures. Elsewhere two of the atoms are meeting full tilt and rebounding, with further disaster to their scanty remains of vesture....
>
> And what is the result of all this bustle? Very little. Unless we have in mind an extremely long stretch of time the general state of the star remains steady."
>
> Sir Arthur S. Eddington: The Internal Constitution of the Stars, 1926

Frequent collisions lead to a state of thermodynamic equilibrium, which is characterized by a temperature, indicative of the energy distribution of the particles. For example, a free ideal gas in thermodynamic equilibrium is described by a Maxwellian velocity (kinetic energy) distribution.

2.1 Local thermodynamic equilibrium

When the average distance travelled by particles between collisions – the *mean free path* – is much smaller than the dimensions of the system, thermodynamic equilibrium is achieved locally, and the system may assume different temperatures at different points. It is thus described by a temperature distribution. If, moreover, the time elapsed between collisions – the *mean free time* – is much shorter than the timescale for change of macroscopic properties, then thermodynamic equilibrium is secured, but the temperature distribution may change with time. Such is the situation in stars.

Equilibrium between matter and radiation can be achieved as well, by "collisions" (interactions) between mass particles and photons. In this case the radiation becomes a *blackbody* radiation, where the energy distribution of the photons is described by the Planck function, and the temperatures of gas and radiation are the same. As we shall see in more detail in the next chapter, the average mean free path of photons in stellar interiors is many orders of magnitude smaller than typical stellar dimensions. Needless to say, the corresponding mean free time of photons is vanishingly small. Consequently, the gas and the radiation may be assumed in thermodynamic equilibrium locally, that is, the gas temperature is the same as the radiation temperature at each point (although the temperature of a star is neither uniform nor constant). This means that the radiation in stellar interiors is very nearly blackbody radiation, described by the Planck function corresponding to the local *unique* temperature. Such a state is known as *local thermodynamic equilibrium* (LTE). It should be stressed that radiation and matter are not always in a state of equilibrium. For example, the solar radiation passing through the Earth's atmosphere does not reach equilibrium with the gas; the radiation temperature is the effective temperature of the Sun, about 6000 K, while the gas temperature, around 300 K, is more than 20 times lower. Similar situations occur in gaseous nebulae that are illuminated by stars embedded in them. There are also mixtures involving more than two temperatures; for example, in an ionized gas, the temperatures of the electrons, the ions, and the photons may all differ from each other. Such is the situation in the solar wind – the flux of particles, mainly protons and electrons, emanating from the Sun. In this case, the characteristic temperatures of the two gases – about 10^6 K for the protons, and almost twice as much for the electrons – are higher than that of the radiation (6000 K).

The assumption of LTE constitutes a great simplification, for it enables the calculation of all thermodynamic properties in terms of the temperature, the density, and the composition, as they change from the stellar centre to the surface. Thus the *structure* of a star of given mass M is uniquely determined at any given time t, if the density ρ, the temperature T, and the composition – the mass fractions of all the constituents – are known at each point within it. By "point" we mean any value of the independent space variable chosen (r or m), which refers to a spherical surface around the centre. The temperature, density, and composition

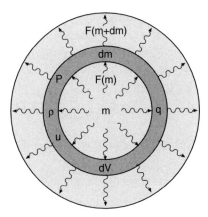

Figure 2.1 Spherical shell within a star and the heat flow into and out of it.

change not only with distance from the centre of the star, but also with time. Hence the *evolution* of a star composed of n different elements is described by the $n+2$ functions: $\rho(m,t)$, $T(m,t)$, and the mass fractions $X_i(m,t)$, where $1 \leq i \leq n$, of two independent variables, time and space. A set of $n+2$ equations is thus required, of which these functions are the solutions.

We thus invoke the basic conservation laws that apply to any physical system: conservation of mass, momentum, angular momentum, and energy. As we have assumed a star to be a nonrotating system, the angular momentum is uniformly zero at all times. (Nevertheless, the global conservation of angular momentum will be invoked later to explain special features of peculiar stars.) Conservation of mass is implicitly included in the relation between dm and dr. Only two conservation laws remain to be applied, for energy and momentum, which together with the equations for the rate of change of abundance for each species will form the set of equations of stellar evolution.

2.2 The energy equation

The first law of thermodynamics, or the principle of conservation of energy, states that the internal energy of a system may be changed by two forms of energy transfer: heat and work. Heat may be added or extracted, and work may be done on the system, or performed by the system, and involves a change in its volume – expansion or contraction. Consider a small element of mass dm within a star, over which the temperature, density, and composition may be taken as approximately constant. In view of the spherical symmetry assumed, such an element may be chosen as a thin spherical shell between radii r and $r+dr$ – as shown in Figure 2.1 – so that its volume is $dV = 4\pi r^2 dr$ and

$$dm = \rho \, dV = \rho \, 4\pi r^2 dr \tag{2.1}$$

[cf. equation (1.5)]. Let u be the internal energy per unit mass and P the pressure. We denote by δf a change that occurs in the value of any quantity f within the mass element over a small period of time δt (a Lagrangian rather than Eulerian change). Then, if δQ is the amount of heat absorbed ($\delta Q > 0$) or emitted ($\delta Q < 0$) by the mass element, and δW is the work done on it during the time interval δt, the change in the internal energy, according to the first law, is given by

$$\delta(udm) = dm\delta u = \delta Q + \delta W, \tag{2.2}$$

where we have used the conservation of mass in assuming dm to be constant. The work may be expressed as

$$\delta W = -P\delta dV = -P\delta\left(\frac{dV}{dm}dm\right) = -P\delta\left(\frac{1}{\rho}\right)dm. \tag{2.3}$$

We note that compression means shrinking of the element's volume, or $\delta dV < 0$, and hence entails an addition of energy, while expansion ($\delta dV > 0$) is achieved at the expense of the element's own energy.

The sources of heat of the mass element are (a) the release of nuclear energy, if available, and (b) the balance of the heat fluxes streaming into the element and out of it. The rate of nuclear energy release per unit mass is denoted by q and the heat flowing perpendicularly through a spherical surface by $F(m)$. Thus F has the dimension of power (not to be confused with the strict definition of a heat flux – power per unit area), and obviously $F(M) \equiv L$. Accordingly,

$$\delta Q = q\,dm\delta t + F(m)\delta t - F(m+dm)\delta t.$$

But $F(m+dm) = F(m) + (\partial F/\partial m)dm$, and hence

$$\delta Q = \left(q - \frac{\partial F}{\partial m}\right)dm\delta t. \tag{2.4}$$

Substituting equations (2.3) and (2.4) in equation (2.2), we may write the latter as

$$dm\delta u + P\delta\left(\frac{1}{\rho}\right)dm = \left(q - \frac{\partial F}{\partial m}\right)dm\delta t, \tag{2.5}$$

and in the limit $\delta t \to 0$ we obtain

$$\dot u + P\left(\frac{1}{\rho}\right)^{\!\cdot} = q - \frac{\partial F}{\partial m}, \tag{2.6}$$

where we have used the notation $\dot f$ for the temporal (partial) derivative $\partial f/\partial t$ of a function f (the notation introduced by Newton).

2.3 The equation of motion

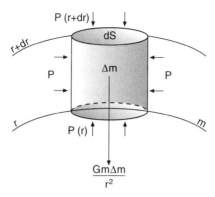

Figure 2.2 Cylindrical mass element within a star.

In *thermal equilibrium*, when temporal derivatives vanish, we have

$$q = \frac{dF}{dm}. \tag{2.7}$$

Integrating over the mass,

$$\int_0^M q\,dm = \int_0^M dF = L, \tag{2.8}$$

for the heat flow must vanish at the centre to avoid singularity. The left-hand side is the total power supplied in the star by nuclear processes, which is commonly denoted by L_{nuc}, the *nuclear luminosity*,

$$L_{\text{nuc}} \equiv \int_0^M q\,dm, \tag{2.9}$$

and thus thermal equilibrium implies that energy is radiated away by the star at the same rate as it is produced in its interior, $L = L_{\text{nuc}}$.

2.3 The equation of motion

Newton's second law of mechanics, or the equation of motion, states that the net force acting on a body of fixed mass imparts to it an acceleration that is equal to the force divided by the mass. This is the momentum conservation law for a body of fixed mass. Consider a small cylindrical volume element within a star, with an axis of length dr in the radial direction, between radii r and $r + dr$, and a cross-sectional area dS, as shown in Figure 2.2. If the (approximately uniform) density within the element is ρ, its mass Δm is given by

$$\Delta m = \rho\,dr\,dS. \tag{2.10}$$

The forces acting on this element are of two kinds: (a) the gravitational force, exerted by the mass of the sphere interior to r (the net gravitational force exerted by the spherical mass shell exterior to r vanishes), and (b) forces resulting from the pressure exerted by the gas surrounding the element. The gravitational force is radial and directed toward the centre of the star. Due to the spherical symmetry assumed, the pressure forces acting perpendicularly to the side of the cylindrical element are balanced and only the pressure forces acting perpendicularly to its top and bottom remain to be considered. Denoting by $\ddot{r}$ the acceleration $\partial^2 r/\partial t^2$ of the element, we may write the equation of motion in the form

$$\ddot{r}\Delta m = -\frac{Gm\Delta m}{r^2} + P(r)\,dS - P(r+dr)\,dS. \tag{2.11}$$

But $P(r+dr) = P(r) + (\partial P/\partial r)dr$ and hence

$$\ddot{r}\Delta m = -\frac{Gm\Delta m}{r^2} - \frac{\partial P}{\partial r}\frac{\Delta m}{\rho},$$

where we have substituted Δm from equation (2.10). We may now divide by Δm to obtain

$$\ddot{r} = -\frac{Gm}{r^2} - \frac{1}{\rho}\frac{\partial P}{\partial r}. \tag{2.12}$$

If m is chosen as the independent space variable rather than r, using the transformation $dr = dm/(4\pi r^2 \rho)$, equation (2.12) becomes

$$\ddot{r} = -\frac{Gm}{r^2} - 4\pi r^2 \frac{\partial P}{\partial m}. \tag{2.13}$$

When accelerations are negligible, equations (2.12) and (2.13) describe a state of *hydrostatic equilibrium*, with gravitational and pressure forces exactly in balance:

$$\frac{dP}{dr} = -\rho\,\frac{Gm}{r^2}, \tag{2.14}$$

or

$$\frac{dP}{dm} = -\frac{Gm}{4\pi r^4}. \tag{2.15}$$

As the right-hand side of equation (2.14) or (2.15) is always negative, hydrostatic equilibrium implies that the pressure decreases outward. The pressure gradient vanishes at the centre, since on the right-hand side of equation (2.14) m/r^2 tends to zero with r.

2.4 The virial theorem

We may estimate the pressure at the centre of a star in hydrostatic equilibrium by integrating equation (2.15) from the centre to the surface of the star,

$$P(M) - P(0) = -\int_0^M \frac{Gm\,dm}{4\pi r^4}. \tag{2.16}$$

On the left-hand side we are left with the central pressure $P_c \equiv P(0)$, since at the surface the pressure practically vanishes, $P(M) \approx 0$. On the right-hand side we may replace r by the stellar radius $R \geq r$, to obtain a lower limit for the central pressure:

$$P_c = \int_0^M \frac{Gm\,dm}{4\pi r^4} > \int_0^M \frac{Gm\,dm}{4\pi R^4}, \tag{2.17}$$

yielding

$$P_c > \frac{GM^2}{8\pi R^4} = 4.4 \times 10^{13} \left(\frac{M}{M_\odot}\right)^2 \left(\frac{R_\odot}{R}\right)^4 \text{ N m}^{-2}. \tag{2.18}$$

The pressure at the centre of the Sun exceeds 450 million atmospheres!

Exercise 2.1: For a star of mass M and radius R, find the central pressure and check the validity of inequality (2.18) for the following cases: (a) a uniform density and (b) a density profile as in Exercise 1.3.

Exercise 2.2: Suppose that the greatest density in a star is ρ_c at the centre and let P_c be the corresponding pressure. Show that

$$P_c < (4\pi)^{1/3} 0.347\, GM^{2/3} \rho_c^{4/3}.$$

2.4 The virial theorem

An important consequence of hydrostatic equilibrium is a link that it establishes between gravitational potential energy and internal energy (or kinetic energy in a system of free particles). Multiplying the equation of hydrostatic equilibrium (2.15) by the volume $V = \frac{4}{3}\pi r^3$ and integrating over the whole star, we obtain

$$\int_0^{P(R)} V\,dP = -\frac{1}{3}\int_0^M \frac{Gm\,dm}{r}. \tag{2.19}$$

The integral on the right-hand side of equation (2.19) is none other than the gravitational potential energy of the star, that is, the energy required to assemble the star by bringing matter from infinity,

$$\Omega = -\int_0^M \frac{Gm\,dm}{r}. \tag{2.20}$$

The left-hand side of equation (2.19) can be integrated by parts:

$$\int_0^{P(R)} V\,dP = [PV]_0^R - \int_0^{V(R)} P\,dV. \tag{2.21}$$

The first term on the right-hand side vanishes, since at the centre $V = 0$ and at the surface $P = 0$. Combining equations (2.19)–(2.21), we finally obtain

$$-3\int_0^{V(R)} P\,dV = \Omega, \tag{2.22}$$

or, since $dV = dm/\rho$,

$$-3\int_0^M \frac{P}{\rho}\,dm = \Omega. \tag{2.23}$$

This is the general, global form of the virial theorem, which will prove extremely valuable in many later discussions. A similar relation, applicable to part of the star, may be obtained by carrying the integration of equation (2.19) up to a radius $R_s < R$:

$$P_s V_s - \int_0^{M_s} \frac{P}{\rho}\,dm = \tfrac{1}{3}\Omega_s. \tag{2.24}$$

Here, Ω_s is the gravitational potential energy of the sphere whose boundary is at R_s, which is unaffected by the shell outside it (between R_s and R), while P_s is the pressure at R_s, exerted by the weight of the enveloping shell.

Consider the particular case of an ideal gas of density ρ and temperature T (to be treated in more detail in the next chapter); let the mass of a gas particle be m_g. The gas pressure is then given by $P = (\rho/m_g)kT$, where k is the Boltzmann constant. The kinetic energy per particle is $\tfrac{3}{2}kT$, and since for an ideal gas the internal energy is the kinetic energy of its particles, the internal energy per unit mass is

$$u = \frac{3}{2}\frac{kT}{m_g} = \frac{3}{2}\frac{P}{\rho}. \tag{2.25}$$

Combining equation (2.25) with the virial theorem (2.23), we have $\int_0^M u\,dm = -\tfrac{1}{2}\Omega$. The integral on the left-hand side is simply the total internal energy U, and

2.4 The virial theorem

hence

$$U = -\tfrac{1}{2}\Omega. \tag{2.26}$$

We can use this result to estimate the average internal temperature of a star (assuming that stellar material behaves as an ideal gas – an assumption that will be justified later on). The gravitational potential energy, equation (2.20), of a star of mass M and radius R is given by

$$\Omega = -\alpha \frac{GM^2}{R}, \tag{2.27}$$

where α is a constant of the order of unity, determined by the distribution of matter within the star, that is, by the density profile. By the virial theorem, we have on the one hand $U = \tfrac{1}{2}\alpha GM^2/R$; on the other hand, from equation (2.25),

$$U = \int_0^M \frac{3}{2}\frac{kT}{m_g} dm = \frac{3}{2}\frac{k}{m_g}\bar{T}M, \tag{2.28}$$

where $\bar{T}$ is the temperature averaged over the stellar mass. Combining the two results, we obtain

$$\bar{T} = \frac{\alpha}{3}\frac{m_g G}{k}\frac{M}{R}. \tag{2.29}$$

Substituting the average density $\bar{\rho} = 3M/4\pi R^3$ in equation (2.29), we obtain $\bar{T} \propto M^{2/3}\bar{\rho}^{1/3}$, meaning that between two stars of the same mass, the denser one is also the hotter.

> **Exercise 2.3:** For a star of mass M and radius R, find the value of α in the expression for the gravitational potential energy for two cases: (a) a uniform density and (b) a density profile as in Exercise 1.3.

Taking $\alpha = \tfrac{1}{2}$ and assuming the gas to be atomic hydrogen, we find that the average temperature of a star is

$$\bar{T} \approx 4 \times 10^6 \left(\frac{M}{M_\odot}\right)\left(\frac{R_\odot}{R}\right) \text{ K}. \tag{2.30}$$

We note that $\bar{T}$ is much higher than the surface temperature T_{eff} (as obtained from observations), implying that internal temperatures must reach still higher values. At temperatures of millions of degrees Kelvin, hydrogen and helium are completely ionized, and even heavier elements are found in gaseous, highly ionized

2 The equations of stellar evolution

form. Stellar material is therefore a *plasma*, a mixture of ions – nuclei stripped of almost all their electrons – and free electrons.

2.5 The total energy of a star

We start by integrating the energy equation (2.6), over the entire star:

$$\int_0^M \dot{u}\, dm + \int_0^M P\left(\frac{1}{\rho}\right)^{\cdot} dm = L_{\text{nuc}} - L. \qquad (2.31)$$

Since the variables t and m are independent, the order of differentiation and integration may be interchanged. Hence the first term on the left-hand side is

$$\int_0^M \dot{u}\, dm = \frac{d}{dt}\int_0^M u\, dm = \dot{U}. \qquad (2.32)$$

Now

$$\left(\frac{1}{\rho}\right)^{\cdot} = \left(\frac{\partial V}{\partial m}\right)^{\cdot} = \frac{\partial \dot{V}}{\partial m} \qquad (2.33)$$

and

$$\dot{V} = 4\pi r^2 \dot{r}. \qquad (2.34)$$

Integrating by parts the second term on the left-hand side of equation (2.31), we obtain

$$\int_0^M P\frac{\partial \dot{V}}{\partial m}dm = [P\dot{V}]_0^M - \int_0^M 4\pi r^2 \dot{r}\frac{\partial P}{\partial m}dm, \qquad (2.35)$$

and since at the centre $\dot{V}$ vanishes with r and at the surface P vanishes, we finally have

$$\dot{U} - \int_0^M 4\pi r^2 \dot{r}\frac{\partial P}{\partial m}dm = L_{\text{nuc}} - L. \qquad (2.36)$$

We turn now to the equation of motion and integrate it, too, over the entire star, after multiplying by $\dot{r}$:

$$\int_0^M \ddot{r}\dot{r}\, dm = -\int_0^M \frac{Gm}{r^2}\dot{r}\, dm - \int_0^M 4\pi r^2 \dot{r}\frac{\partial P}{\partial m}dm. \qquad (2.37)$$

2.5 The total energy of a star

As the total kinetic energy of the star is given by

$$\mathcal{K} = \int_0^M \tfrac{1}{2}\dot{r}^2 dm, \tag{2.38}$$

the integral on the left-hand side of equation (2.37) is

$$\int_0^M \dot{r}\ddot{r}\, dm = \int_0^M \frac{\partial}{\partial t}(\tfrac{1}{2}\dot{r}^2)dm = \frac{d}{dt}\int_0^M \tfrac{1}{2}\dot{r}^2 dm = \dot{\mathcal{K}}. \tag{2.39}$$

The first term on the right-hand side of equation (2.37) is

$$-\int_0^M Gm\frac{\dot{r}}{r^2}dm = \int_0^M Gm\left(\frac{1}{r}\right)^{\cdot} dm = \frac{d}{dt}\int_0^M \frac{Gm\,dm}{r} = -\dot{\Omega}. \tag{2.40}$$

Thus equation (2.37) reads

$$\dot{\mathcal{K}} + \dot{\Omega} = -\int_0^M 4\pi r^2 \dot{r}\frac{\partial P}{\partial m}dm. \tag{2.41}$$

Combining equations (2.36) and (2.41), we have

$$\dot{U} + \dot{\mathcal{K}} + \dot{\Omega} = L_{\text{nuc}} - L, \tag{2.42}$$

where the left-hand side is the rate of change of the total stellar energy, that is, $E = U + \mathcal{K} + \Omega$,

$$\dot{E} = L_{\text{nuc}} - L. \tag{2.43}$$

If a star is in thermal equilibrium, it follows that $\dot{E} = 0$ and the energy is constant. If, in addition, the star is in hydrostatic equilibrium, $\mathcal{K}$ vanishes. In this case U and Ω are related by the virial theorem, and hence either of them determines the total energy of the star. Consequently, each one of them is conserved, not only their sum. For example, a star in thermal and hydrostatic equilibrium cannot cool throughout and expand (although cooling decreases the energy and expansion increases it); it must conserve the internal (thermal) energy and the gravitational potential energy separately. Another, apparently puzzling, conclusion is that a star in hydrostatic equilibrium has a negative heat capacity, meaning that it becomes hotter upon losing energy! This follows again from the virial theorem: for an ideal gas, we have

$$E = U + \Omega = \tfrac{1}{2}\Omega = -U, \tag{2.44}$$

and in the general case the right-hand side is multiplied by a constant of the order of unity. Hence if $\dot{E} < 0$, then U, and with it [by equation (2.28)] the average temperature of the star, must increase. At the same time, the star must contract. (We shall see shortly that contraction does not necessarily imply violation of hydrostatic equilibrium, so that the last argument is not contradictory.) In fact, the gravitational potential energy released in contraction supplies both the energy that is lost (radiated) *and* the thermal energy that causes the temperature to rise – in equal amounts in the case of an ideal gas.

> **Exercise 2.4:** Assuming that a star of mass M is devoid of nuclear energy sources, find the rate of contraction of its radius, if it maintains a constant luminosity L.

2.6 The equations governing composition changes

As we have seen that stellar material is composed of free electrons and (almost) entirely bare, chemically unbound nuclei, composition changes – if any – cannot be of a chemical nature. The only possible changes in the abundance of the constituents can occur by transformations from one element into another, that is, by nuclear reactions – interactions between nuclei.

The atomic nucleus is made of protons and neutrons (collectively called *nucleons*), which belong to the class of *heavy particles* named *baryons* ("baryon" meaning "heavy one" in Greek). The proton has a positive electric charge of unity (in units of the elementary charge e); the neutron has zero charge. These particles may therefore be characterized by two numbers, baryon number $\mathcal{A}$ and charge $\mathcal{Z}$, $(1, +1)$ for the proton and $(1,0)$ for the neutron. Electrons, like protons, are charged particles. Their baryon number is 0, meaning that electrons are not heavy particles (indeed, the electron mass is almost 2,000 times smaller than the proton mass). Associated with the electrons are the *neutrinos*, of baryon number 0 (the neutrino mass is still controversial) and charge 0. Electrons and neutrinos belong to a class of particles called *leptons* ("light ones" in Greek). For each particle, relativistic quantum mechanics postulates the existence of an *antiparticle*, for which the signs of baryon or lepton number and charge are reversed. The best known antiparticle is the *positron*, the electron's antiparticle.

Protons and neutrons are bound together in the atomic nucleus by a force of attraction called *the strong nuclear force*. This is a short-range force, independent of charge, that surpasses the repulsive Coulomb force between protons at nuclear length scales – a few *fermis* (1 fermi = 10^{-15} m). Another force that can act on protons and neutrons is the *weak force*, whose range is estimated to be still shorter ($<10^{-17}$ m). The weak interaction is responsible for the conversion of protons into neutrons (or vice versa). In a nuclear reaction (interaction by means of the strong or the weak force) the charge as well as the baryon and lepton numbers are conserved. Hence in a weak interaction, an electron or a positron must be involved in order to conserve charge, and a neutrino or antineutrino, so as to conserve the

2.6 The equations governing composition changes

lepton number. Conservation of lepton number means equal numbers of leptons and antileptons; hence a positron (antilepton) will be accompanied by a neutrino and an electron by an antineutrino.

If the bulk density in some part of a star is ρ and the partial density of the ith nuclear species is ρ_i, the mass fraction of this species is given by

$$X_i = \frac{\rho_i}{\rho}, \tag{2.45}$$

and the number density – number of nuclei per unit volume – is given by the partial density divided by the mass of one nucleus. The mass of an atomic nucleus is slightly less than the sum of masses of its constituent protons and neutrons as free particles. However, to a good approximation we may write

$$n_i = \frac{\rho_i}{\mathcal{A}_i m_H}, \tag{2.46}$$

where m_H is the *atomic mass unit* representing the mass of a (bound) nucleon, usually defined as a twelfth of a carbon nucleus mass. Despite the notation, m_H is slightly different both from the proton mass and from the mass of a hydrogen atom. Combining equations (2.45) and (2.46), we obtain the relations

$$n_i = \frac{\rho}{m_H} \frac{X_i}{\mathcal{A}_i} \quad \text{and} \quad X_i = n_i \frac{\mathcal{A}_i}{\rho} m_H. \tag{2.47}$$

The number of nuclei in a given volume may change as a result of nuclear reactions that create it and others that destroy it. The creation or destruction of a nucleus takes place by fusion of lighter nuclei or by breakup of a heavier nucleus and may involve capture and release of light particles, such as positrons and electrons, neutrinos (and antineutrinos), and energetic photons. Specific nuclear reactions will be considered in Chapter 4. A general way of describing a nuclear reaction is by two different nuclei combining to produce two other nuclei. Since the nucleus of any element is uniquely defined by the two integers $\mathcal{A}_i$ and $\mathcal{Z}_i$, we denote the reactants by the symbols $I(\mathcal{A}_i, \mathcal{Z}_i)$ and $J(\mathcal{A}_j, \mathcal{Z}_j)$, and the products by the symbols $K(\mathcal{A}_k, \mathcal{Z}_k)$ and $L(\mathcal{A}_l, \mathcal{Z}_l)$. A nuclear reaction can proceed in either direction (similarly to chemical reactions and ionization-recombination processes), depending on the temperature (kinetic energy) and density of the particles, and can therefore be described symbolically by

$$I(\mathcal{A}_i, \mathcal{Z}_i) + J(\mathcal{A}_j, \mathcal{Z}_j) \rightleftharpoons K(\mathcal{A}_k, \mathcal{Z}_k) + L(\mathcal{A}_l, \mathcal{Z}_l), \tag{2.48}$$

subject to two conservation laws:

$$\mathcal{A}_i + \mathcal{A}_j = \mathcal{A}_k + \mathcal{A}_l, \tag{2.49}$$

$$\mathcal{Z}_i + \mathcal{Z}_j = \mathcal{Z}_k + \mathcal{Z}_l. \tag{2.50}$$

If positrons or electrons are also involved, they must be taken into account as well: for them $\mathcal{A} = 0$ and $\mathcal{Z} = \pm 1$ and conservation of lepton number must be obeyed. Therefore, any three of the four nuclei involved in the reaction uniquely determine the fourth. The reaction rate, say from left to right, can be identified by three indices: two for the reactants and one for one of the products.

Let us now attempt to evaluate the rate at which nuclei of type I are destroyed by reactions of type (2.48) with the help of a simplified picture. Consider a unit volume and assume that each I nucleus within it has a cross-sectional area ς, meaning that any J nucleus striking this area will cause a reaction to occur. Assume further that the relative velocity of I nuclei with respect to J nuclei is v, so that I nuclei may be considered as targets at rest, while J nuclei flow toward them at velocity v. The effective target area is therefore $n_i \varsigma$; the number of particles crossing a unit area per unit time is $n_j v$. Hence the number of reactions that occur per unit time in this unit volume is $n_i n_j \varsigma v$ or $n_i n_j R_{ijk}$, where $R_{ijk} \equiv \varsigma v$ – having the dimension of volume divided by time – is called the *reaction rate*. In the case of particles of one kind, say I, interacting with each other, the product $n_i n_j$ is replaced by $\frac{1}{2} n_i^2$. The velocity of gas particles in the star is the thermal velocity – this is why nuclear reactions occurring in stars are called *thermonuclear reactions* – and the cross-section depends on the properties of the nuclei involved (such as their charges) as well as on the properties of the products.

We may now write the rate of change of the ith element's abundance resulting from all possible nuclear reactions, both destructive and constructive, in the form

$$\dot{n}_i = -n_i \sum_{j,k}(1+\delta_{ij})\frac{n_j}{1+\delta_{ij}} R_{ijk} + \sum_{l,k} \frac{n_l n_k}{1+\delta_{lk}} R_{lki}, \qquad (2.51)$$

noting that two particles of type i are destroyed when $j = i$. Alternatively, ...

$$\frac{\dot{X}_i}{A_i} = \frac{\rho}{m_H}\left(-\frac{X_i}{A_i}\sum_{j,k}(1+\delta_{ij})\frac{X_j}{A_j}\frac{R_{ijk}}{1+\delta_{ij}} + \sum_{l,k}\frac{X_l}{A_l}\frac{X_k}{A_k}\frac{R_{lki}}{1+\delta_{lk}}\right) \qquad (2.52)$$

and similar equations for the other mass fractions. For simplicity, we may define a composition *vector* by $\mathbf{X} \equiv (X_1, \ldots, X_n)$ so that the set of equations describing composition changes may be symbolically written as one equation with n components, one for each element,

$$\dot{\mathbf{X}} = \mathbf{f}(\rho, T, \mathbf{X}). \qquad (2.53)$$

In *nuclear equilibrium*, when $\dot{X}_i = 0$, X_i is readily obtained from $f_i = 0$.

2.7 The set of evolution equations

The set of nonlinear partial differential equations describing the evolutionary course of the internal structure of a star is

$$\ddot{r} = -\frac{Gm}{r^2} - 4\pi r^2 \frac{\partial P}{\partial m},$$

$$\dot{u} + P\left(\frac{1}{\rho}\right)^{\cdot} = q - \frac{\partial F}{\partial m}, \qquad (2.54)$$

$$\dot{\mathbf{X}} = \mathbf{f}(\rho, T, \mathbf{X}).$$

As it stands, the set is not complete; besides the structure functions – $\rho(m,t)$, $T(m,t)$, and $X_i(m,t)$ – that form the set of unknowns, it contains additional functions: (a) P and u, (b) F, and (c) q and $\mathbf{f}$, which have to be supplied in terms of the unknowns. To this purpose, we shall have to invoke different branches of physics: thermodynamics and statistical mechanics in case (a); atomic physics and the theory of radiation transfer in case (b); and nuclear and elementary particle physics in case (c). This is, in fact, what distinguishes astrophysics from other physical disciplines. Astrophysics does not deal with a special, distinct class of effects and processes, as do the basic fields of physics. Nuclear physics, for example, deals exclusively with the atomic nucleus; there are many ramifications to this field of research, such as nuclear forces, nuclear structure, and nuclear reactions, but they are all intimately connected. Nuclear physics has very little to do, say, with hydrodynamics, the study of the motion of continuous media. By contrast, astrophysics deals with complex phenomena, which involve processes of many different kinds. It has to lean, therefore, on all the branches of physics, and this makes for its special beauty. The theory of the structure and evolution of stars presents a unique opportunity to bring separate, seemingly unconnected, physical theories under one roof. In the next chapter we shall interrupt our pursuit of the evolutionary course of stars, in order to extract from different physical theories the information that will enable us to resume it.

Finally, in order for the set of differential equations to be solved, boundary and initial conditions have to be supplied. The two space derivatives require two boundary conditions, and the $n+3$ time derivatives require $n+3$ initial distributions of physical properties. The boundary conditions are straightforward: $P(M,t) = 0$ and, in order to avoid a singularity at the centre, $F(0,t) = 0$. The initial conditions could be $\rho(m,0)$, $T(m,0)$, $\dot{r}(m,0)$, and $X_i(m,0)$, or related functions. Here, it seems, we run into serious difficulties; as mentioned before, star formation is still a subject of study. The initial state of a star is, therefore, rather obscure. Fortunately, as we shall see shortly, this problem can be overcome, or more precisely, avoided.

2.8 The characteristic timescales of stellar evolution

The evolution of a star is described by the three time-dependent equations (2.54), each dealing with a different type of change: the first involves dynamical or structural changes, the second describes thermal changes, and the third deals with nuclear processes leading to changes in composition (and in the rest-mass energy). Each change, or process, has its characteristic timescale τ, which can be defined as the ratio of the quantity (or physical property) ϕ that is changed by the process and the rate of change of this quantity:

$$\tau = \frac{\phi}{\dot{\phi}}. \tag{2.55}$$

The simplest example would be of a motion, whose duration – or characteristic timescale – is given by distance divided by velocity. Obviously, a rapid process (large $\dot{\phi}$) has a short timescale and *vice versa*. It is instructive to estimate and compare the timescales of the different processes that occur in stars.

The dynamical timescale

We can envisage a considerable change in the structure of a spherically symmetric star as a change in its characteristic dimension, the radius R; hence in this case we may take $\phi = R$. As gravity is the binding force of a star, the typical rate of change of R would be the characteristic velocity in a gravitational field: the free fall or escape velocity $v_{\text{esc}} = \sqrt{2GM/R}$; hence $\dot{\phi} = \sqrt{2GM/R}$. The dynamical timescale may therefore be estimated by

$$\tau_{\text{dyn}} \approx \frac{R}{v_{\text{esc}}} = \sqrt{\frac{R^3}{2GM}}, \tag{2.56}$$

or, in terms of the average density $\bar{\rho} = 3M/4\pi R^3$, neglecting factors of the order of unity,

$$\tau_{\text{dyn}} \approx \frac{1}{\sqrt{G\bar{\rho}}}. \tag{2.57}$$

There are many ways to obtain the dynamical timescale, but they all lead to the same result, within factors of the order of unity. The dynamical timescale of the Sun is about 1000 s (roughly a quarter of an hour), and generally

$$\tau_{\text{dyn}} \approx 1000 \sqrt{\left(\frac{R}{R_\odot}\right)^3 \left(\frac{M_\odot}{M}\right)} \text{ s.} \tag{2.58}$$

2.8 The characteristic timescales of stellar evolution

The dynamical timescale is extremely short, many orders of magnitude shorter than typical stellar ages. The estimated age of the Sun, for example, is 4.6 billion years, or $\sim 1.5 \times 10^{17}$ s, about $10^{14}\tau_{\rm dyn}$. What is the meaning of this result? A dynamical process occurs in a star whenever the gravitational force is not balanced by the pressure forces [see equation (2.12)]. Such a situation can develop either into contraction, if there is insufficient pressure to counteract gravity, or into expansion, if the pressure is too high. It can end either in a catastrophic event – collapse or explosion – or in restoration of hydrostatic equilibrium, when the forces are again in balance. Either of these end states will be achieved within a period of time of the order of the dynamical timescale. This leads us to the following conclusions.

1. If a star cannot recover from a dynamical process (by restoring hydrostatic equilibrium), the ensuing collapse – or explosion – should be observable in its entirety. Indeed, such events have been known to occur: they are called *supernovae*. We shall return to them in Chapter 9.
2. Rapid changes that are sometimes observed in stars may indicate that dynamical processes are taking place, but on a smaller scale, not involving the entire star. From the timescale of such changes – usually oscillations with a characteristic period – we may roughly estimate the average density of the star. The Sun has been observed to oscillate with a period of minutes. Oscillations with periods of a few tens of seconds indicate that the star should be a compact one, such as a white dwarf.
3. As a rule, stars may be assumed to be in a state of hydrostatic equilibrium throughout. Any perturbation of this state is immediately quenched. This does not mean, of course, that stars are static during their entire life span, but rather that they evolve *quasi-statically*, constantly adjusting their internal structure so as to maintain dynamical balance. Consequently, the left-hand side of equation (2.12) may be assumed to vanish, and the virial theorem (Section 2.4) may be assumed to hold at all times. This means that the gravitational potential energy and the thermal energy of the star each follows the behaviour of the total energy.

The thermal timescale

Thermal processes affect the internal energy of the star; hence in this case we may take $\phi = U$. By the virial theorem (which, as we have seen, is applicable), $U \approx GM^2/R$. The characteristic rate of change of U is the rate at which energy is radiated away by the star; thus we may set $\dot\phi = L$. The thermal timescale may be therefore estimated by

$$\tau_{\rm th} = \frac{U}{L} \approx \frac{GM^2}{RL}. \qquad (2.59)$$

For the Sun, $\tau_{th} \approx 10^{15}$ s, or about 30 million years, and generally

$$\tau_{th} \approx 10^{15} \left(\frac{M}{M_\odot}\right)^2 \left(\frac{R_\odot}{R}\right)\left(\frac{L_\odot}{L}\right) \quad \text{s}. \tag{2.60}$$

The thermal timescale is many orders of magnitude longer than the dynamical timescale, but it still constitutes only a small fraction – about 1% or less – of the life span of a star. Thus, although we would not be able to observe the development of a thermal process in a star (in fact, we have no way of knowing whether any observed star is in thermal equilibrium or not), we may assume that throughout most of its life a star is in a state of thermal equilibrium. If a star maintains both thermal and hydrostatic equilibrium during an evolutionary phase, its total energy is conserved (or changes very slowly) during that phase, and by the virial theorem, the gravitational potential energy and the thermal energy, each, is conserved. Thus, if contraction occurs (quasi-statically) in some part of the star, it follows that other parts should expand so as to conserve Ω. Similarly, if the temperature rises in some place, it should decrease in another, so as to keep U constant. Later on we shall make use of such arguments.

The thermal timescale may be interpreted as the time it would take a star to emit its entire reserve of thermal energy upon contracting (as we have shown in Section 2.5), provided it maintains a constant luminosity. This was, in fact, the way William Thomson (better known as Lord Kelvin) and, independently, Hermann von Helmholtz estimated the Sun's age more than a century ago, and for this reason, the thermal timescale is often called the *Kelvin–Helmholtz timescale*.

Historical Note: Kelvin's (1862) estimate imposed an upper limit on the age of the Earth, which was in marked conflict with the new theory put forward by Charles Darwin (in 1859). This theory required that geological time be much longer, so as to account for the slow evolution of countless species of plants and animals (living and fossil) by natural selection. A long and intense debate ensued between the two eminent scientists. To the end Darwin remained convinced that, in time, physicists will change their minds. Harsh criticism of Kelvin's estimate came toward the end of the nineteenth century from the geologist Thomas C. Chamberlin:

"Is present knowledge relative to the behaviour of matter under such extraordinary conditions as obtain in the interior of the sun sufficiently exhaustive to warrant the assertion that no unrecognized sources of heat reside there? What the internal constituents of the atoms may be is yet an open question. It is not improbable that they are complex organizations and the seats of enormous energies."

T. C. Chamberlin: Annual Report of the Smithsonian Institution (1899)

And two decades later Eddington, addressing the same issue, predicted that the source of energy in stars should be "subatomic":

> "Only the inertia of tradition keeps the contraction hypothesis alive – or rather, not alive, but an unburied corpse ...
> A star is drawing on some vast reservoir of energy by means unknown to us. This reservoir can scarcely be other than the subatomic energy which, it is known, exists abundantly in all matter ... There is sufficient in the Sun to maintain its output of heat for 15 billion years ...
> If, indeed, the subatomic energy in the stars is being freely used to maintain their great furnaces, it seems to bring a little nearer to fulfillment our dream of controlling this latent power for the well-being of the human race – or for its suicide."
>
> Sir Arthur S. Eddington: Observatory 43 (1920)

Finally, after about ten more years, the controversy was settled (in Darwin's favour!) by quantum and nuclear physics, which solved the puzzle of the energy source of stars.

The nuclear timescale

The quantity that is changed by nuclear processes, besides abundances, is a (small) fraction of the rest-mass energy given by Einstein's famous relation $E = mc^2$. This fraction, which may be turned into other forms of energy, constitutes the nuclear potential energy. Hence we may take $\phi = \varepsilon M c^2$, where ε can be estimated by the typical binding energy of a nucleon divided by the nucleon's rest-mass energy, which amounts to a few 10^{-3}. The rate of change of the nuclear potential energy is, obviously, the nuclear luminosity L_{nuc}, and since we are allowed to assume thermal equilibrium, we may take $\dot{\phi} = L_{\text{nuc}} = L$. Hence

$$\tau_{\text{nuc}} \approx \frac{\varepsilon M c^2}{L}, \qquad (2.61)$$

or, using solar units,

$$\tau_{\text{nuc}} \approx \varepsilon 4.5 \times 10^{20} \left(\frac{M}{M_\odot}\right)\left(\frac{L_\odot}{L}\right) \text{ s}. \qquad (2.62)$$

For the Sun, this is many times its age; in fact, τ_{nuc} is larger than the estimated age of the universe. An immediate conclusion that emerges is that stars seem to have actually consumed only a small fraction of their available nuclear energy, meaning that only a fraction of the stellar mass has changed its initial composition. Another is that, generally, nuclear equilibrium is not to be expected.

To summarize our results,

$$\tau_{\text{dyn}} \ll \tau_{\text{th}} \ll \tau_{\text{nuc}}. \qquad (2.63)$$

Consequently, it is the rates of nuclear processes that determine the pace of stellar evolution, throughout which the star may be assumed to be in thermal and hydrostatic equilibrium at each stage. The set of evolution equations (2.54) reduces to

$$\boxed{\begin{aligned} \frac{dP}{dm} &= -\frac{Gm}{4\pi r^4}, \\ \frac{dF}{dm} &= q, \\ \dot{\mathbf{X}} &= \mathbf{f}(\rho, T, \mathbf{X}). \end{aligned}} \quad (2.64)$$

This is a considerable simplification of the original set. In particular, we need not know the initial structure of the star in order to be able to trace its evolution. All we need to know is the initial composition, which we shall assume to be homogeneously distributed throughout the star, and this is precisely what we do know reasonably well. Our task of investigating the evolution of stars now divides into two different parts, or two main questions: (a) What is the sequence of nuclear processes that take place in stellar interiors? (b) Given the composition, what is the structure – distribution of temperature and density – of a star in hydrostatic and thermal equilibrium? Clearly, these questions cannot be separated: nuclear processes depend on temperature and density and the structure of a star depends on its composition. But they can be answered in turn and the answers may then be combined into a comprehensive picture of stellar evolution.

We shall address the first question in Chapter 4 and the second in Chapters 5 and 6. The next chapter, dealing in a general manner with the physics of stellar interiors, may be skipped by readers who are familiar with the physics of gaseous systems and of radiation.

3

Elementary physics of gas and radiation in stellar interiors

As a star consists of a mixture of ions, electrons, and photons, the physics of stellar interiors must deal with (a) the properties of gaseous systems, (b) radiation, and (c) the interaction between gas and radiation. The latter may take many different forms: absorption, resulting in excitation or ionization; emission, resulting in de-excitation or recombination; and scattering. In order not to stray too far from our main theme, we shall only consider processes and properties that are simple enough to understand without requiring an extended physical background, and yet sufficient for providing some insight into the general behaviour of stars. The full-scale processes are incorporated in calculations of stellar structure and evolution, performed on powerful computers by means of extended numerical codes that include enormous amounts of information. These, however, should be regarded as computational laboratories, meant to reproduce, or simulate, rather than explain, the behaviour of stars. Our purpose is to outline the basic principles of stellar evolution and we are therefore entitled to some simplification. Eddington defends this right quite forcefully:

> "I conceive that the chief aim of the physicist in discussing a theoretical problem is to obtain "insight" – to see which of the numerous factors are particularly concerned in any effect and how they work together to give it. For this purpose a legitimate approximation is not just an unavoidable evil; it is a discernment that certain factors – certain complications of the problem – do not contribute appreciably to the result. We satisfy ourselves that they may be left aside; and the mechanism stands out more clearly, freed from these irrelevancies. This discernment is only a continuation of a task begun by the physicist before the mathematical premises of the problem could be stated; for in any natural problem the actual conditions are of extreme complexity and the first step is to select those which have an essential influence on the result – in short, to get hold of the right end of the stick. The correct use of this insight, whether before or after the mathematical problem has been formulated, is a faculty to be cultivated, not a vicious propensity to be hidden from the public eye. Needless to say the physicist

must if challenged be prepared to defend the use of his discernment; but unless the defence involves some subtle point of difficulty it may well be left until the challenge is made."

Sir Arthur S. Eddington: The Internal Constitution of the Stars, 1926

3.1 The equation of state

A relation between the pressure exerted by a system of particles of known composition and the ambient temperature and density, $P = P(\rho, T, \mathbf{X})$, is called an *equation of state*.

In previous sections we have repeatedly assumed the stellar gas to be an ideal gas, which implies a mixture of free, noninteracting particles (a *perfect* gas). The time has come to justify this assumption. At the temperatures prevailing in stars, gases are ionized, and Coulomb interactions can be expected to occur. We shall show that the energy involved in such interactions is small compared with the kinetic (thermal) energy of the particles. For an average density $\bar{\rho}$ and a gas particle mass $\mathcal{A} m_H$, the mean interparticle distance is

$$d = \left(\frac{\mathcal{A} m_H}{\bar{\rho}}\right)^{1/3} = \left(\frac{4\pi \mathcal{A} m_H}{3M}\right)^{1/3} R, \qquad (3.1)$$

where $\bar{\rho}$ has been expressed in terms of stellar mass M and radius R. If the particle charge is $\mathcal{Z} e$ (e denoting the electron charge), the typical Coulomb energy per particle may be estimated as

$$\epsilon_C \approx \frac{1}{4\pi \varepsilon_0} \frac{\mathcal{Z}^2 e^2}{d}. \qquad (3.2)$$

The kinetic energy (per particle) is of the order $k\bar{T}$ and hence, after substituting $\bar{T}$ from equation (2.29),

$$\frac{\epsilon_C}{k\bar{T}} \sim \frac{1}{4\pi \varepsilon_0} \frac{\mathcal{Z}^2 e^2}{\mathcal{A}^{4/3} m_H^{4/3} G M^{2/3}}, \qquad (3.3)$$

ignoring factors of the order of unity. For $\mathcal{Z} = 1$, $\mathcal{A} = 1$, and $M = M_\odot$, this ratio is about 1% and, to this accuracy, Coulomb interactions may therefore be neglected. For higher $\mathcal{Z}$, we have $\mathcal{A} \approx 2\mathcal{Z}$, and the ratio (3.3), which varies as $\mathcal{Z}^{2/3}$, remains well below unity even for a composition of pure iron. We should note, however, that $\epsilon_C/k\bar{T} \gtrsim 1$ for mass $M \lesssim 10^{-3} M_\odot$. Although stars do not belong to this dangerous zone, planets do: the mass of Jupiter is roughly $10^{-3} M_\odot$. Consequently, the structure of planets cannot be described by a mixture of free gases; more complicated equations of state must be invoked. Since $\epsilon_C/k\bar{T} \gg 1$

3.1 The equation of state

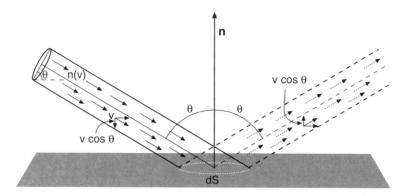

Figure 3.1 Beam of particles impinging on a hypothetical surface, making an angle θ with the normal to the surface.

characterizes solids, we may conclude that the smaller a planet's mass, the closer to a solid is its structure.

A general theorem enables the calculation of the pressure of a free particle system by means of an integral known as the *pressure integral*:

$$P = \tfrac{1}{3} \int_0^\infty v\, p\, n(p)\, dp, \qquad (3.4)$$

where v is the particle velocity, p is its momentum, and $n(p)\, dp$ is the number of particles per unit volume with momenta within the interval $(p, p + dp)$. The proof of this theorem is as follows: Consider a surface (real or imaginary) within the system of particles. The pressure on this surface results from the momentum imparted by particles colliding elastically with it. The momentum transferred by an incident particle is twice the momentum component normal to the surface

$$\Delta p = 2p \cos\theta. \qquad (3.5)$$

Consider a beam of particles impinging on the surface at a velocity v making an angle θ with the normal to the surface (as shown in Figure 3.1). Let $n(\theta, p)\, d\theta\, dp$ be the number density of particles with momenta in the range $(p, p + dp)$ and directions within a cone $(\theta, \theta + d\theta)$. Since the particle distribution is isotropic, the number of particles within any solid angle $d\omega$ is proportional to the solid angle, or

$$\frac{n(\theta, p)\, d\theta\, dp}{n(p)\, dp} = \frac{d\omega}{4\pi} = \frac{2\pi \sin\theta\, d\theta}{4\pi} = \tfrac{1}{2} \sin\theta\, d\theta. \qquad (3.6)$$

The number of particles from this beam striking the surface in a time interval δt is given by the number density multiplied by the volume $v\delta t\, dS \cos\theta$, where dS is the area of incidence of the beam on the surface. Hence the momentum transferred to the surface by these particles is given by

$$\delta p_\theta = n(\theta, p)\, d\theta\, dp\, v\, \delta t\, dS\, \cos\theta\, \Delta p. \tag{3.7}$$

The contribution of these particles to the pressure – the momentum transferred per unit time per unit surface area – is therefore given by equation (3.7), combined with equations (3.5) and (3.6), after dividing by $\delta t\, dS$:

$$dP = \tfrac{1}{2} \sin\theta\, d\theta\, n(p)\, dp\, v \cos\theta\, 2p \cos\theta = vpn(p)\, dp \cos^2\theta \sin\theta\, d\theta, \tag{3.8}$$

and the total pressure is obtained by integrating over all angles of incidence ($0 \leq \theta \leq \pi/2$) and all momenta. Since

$$\int_0^{\pi/2} \cos^2\theta \sin\theta\, d\theta = \int_0^1 \cos^2\theta\, d\cos\theta = \tfrac{1}{3}, \tag{3.9}$$

the proof is completed. Obviously, the pressure of a mixture of free, noninteracting particles of different species will be given by the sum of the pressures exerted by each species separately. This will include the radiation pressure, which is the pressure exerted by photons. Consequently, in stars, P will be the sum of three different terms: P_I for the ions, P_e for the electrons, and P_{rad} for the photons, the first two constituting the total gas pressure,

$$P = P_I + P_e + P_{rad} = P_{gas} + P_{rad}. \tag{3.10}$$

It is customary to define a parameter β as the fraction of the pressure contributed by the gas; thus

$$P_{gas} = \beta P \tag{3.11}$$

$$P_{rad} = (1 - \beta)P. \tag{3.12}$$

3.2 The ion pressure

The equation of state for an ideal ion gas is the well-known relation

$$P_I = n_I kT, \tag{3.13}$$

where n_I is the number of ions per unit volume. This relation is obtained by applying the theorem just proven to a free particle gas in thermodynamic equilibrium, which is characterized by a Maxwellian velocity distribution:

$$n(p)\, dp = \frac{n_I 4\pi p^2\, dp}{(2\pi m_I kT)^{3/2}} e^{-\frac{p^2}{2m_I kT}}. \tag{3.14}$$

Using relations (2.47), we obtain the total number of ions in a unit volume by summing over all the ion species i:

$$n_\mathrm{I} = \sum_i n_i = \sum_i \frac{\rho}{m_\mathrm{H}} \frac{X_i}{\mathcal{A}_i}. \qquad (3.15)$$

The mean atomic mass of stellar material μ_I is defined by

$$\frac{1}{\mu_\mathrm{I}} \equiv \sum_i \frac{X_i}{\mathcal{A}_i}, \qquad (3.16)$$

so that

$$n_\mathrm{I} = \frac{\rho}{\mu_\mathrm{I} m_\mathrm{H}}, \qquad (3.17)$$

and μ_I may be approximated by

$$\frac{1}{\mu_\mathrm{I}} \approx X + \frac{1}{4}Y + \frac{1-X-Y}{\langle \mathcal{A} \rangle}, \qquad (3.18)$$

where $\langle \mathcal{A} \rangle$ is the average atomic mass of the *heavy elements* (elements other than hydrogen and helium, sometimes referred to as *metals*). For the Sun, for example, $X = 0.707$, $Y = 0.274$, and $\langle \mathcal{A} \rangle \approx 20$; hence $\mu_\mathrm{I} = 1.29$. The ratio k/m_H is usually known as the *ideal gas constant*

$$\mathcal{R} \equiv \frac{k}{m_\mathrm{H}}. \qquad (3.19)$$

Substituting equations (3.17) and (3.19) into equation (3.13), we finally obtain

$$P_\mathrm{I} = \frac{\mathcal{R}}{\mu_\mathrm{I}} \rho T. \qquad (3.20)$$

3.3 The electron pressure

If the electrons constitute an ideal gas, the equation of state is, as equation (3.13) above,

$$P_\mathrm{e} = n_\mathrm{e} kT, \qquad (3.21)$$

where n_e is the number of (free) electrons per unit volume. Here we shall make a simplifying assumption by taking the atoms to be completely ionized. This is certainly correct for the main stellar constituents, hydrogen and helium, at temperatures exceeding 10^6 K. The assumption is obviously incorrect for stellar photospheres, but we are mainly concerned with the interior. With this assumption, the total number of electrons per unit volume is

$$n_e = \sum_i Z_i n_i = \frac{\rho}{m_H} \sum_i X_i \frac{Z_i}{A_i}. \tag{3.22}$$

We define μ_e^{-1} as the average number of free electrons per nucleon,

$$\frac{1}{\mu_e} \equiv \sum_i X_i \frac{Z_i}{A_i}, \tag{3.23}$$

leading to

$$n_e = \frac{\rho}{\mu_e m_H}. \tag{3.24}$$

In terms of the mass fractions X and Y, we have

$$\frac{1}{\mu_e} = X + \tfrac{1}{2} Y + (1 - X - Y) \left\langle \frac{Z}{A} \right\rangle, \tag{3.25}$$

where $\langle \frac{Z}{A} \rangle$ is the average value for *metals*, which may be approximated reasonably well by $\tfrac{1}{2}$. Hence

$$\frac{1}{\mu_e} \approx \tfrac{1}{2}(1 + X), \tag{3.26}$$

which for the Sun amounts to $\mu_e \approx 1.17$ and for hydrogen-depleted stars to $\mu_e \approx 2$. The electron pressure is thus given by

$$P_e = \frac{\mathcal{R}}{\mu_e} \rho T. \tag{3.27}$$

Combining equations (3.20) and (3.27), we obtain the total gas pressure

$$P_{\text{gas}} = P_I + P_e = \left(\frac{1}{\mu_I} + \frac{1}{\mu_e} \right) \mathcal{R} \rho T = \frac{\mathcal{R}}{\mu} \rho T, \tag{3.28}$$

where

$$\frac{1}{\mu} \equiv \frac{1}{\mu_I} + \frac{1}{\mu_e}, \tag{3.29}$$

yielding $\mu = 0.61$ for the solar composition. Note that for hydrogen, the contributions of ions and electrons to the gas pressure are equal; for all heavier elements, the electron pressure is higher than the ion pressure (twice as high for helium, for example).

3.3 The electron pressure

The assumptions explicitly made so far were (a) lack of interactions between gas particles, and (b) complete ionization. Other assumptions were, however, implicitly included when adopting the classical physics approach, ignoring quantum and relativistic effects. But the conditions of stellar interiors are such that these effects cannot always be neglected.

According to quantum mechanics, the simultaneous position and momentum of an electron (or any other particle) cannot be known more precisely than allowed by the *Heisenberg uncertainty principle*. More specifically, if a particle's location is known to be within a volume element ΔV and its momentum is within an element $\Delta^3 p$ in the three-dimensional momentum space, then ΔV and $\Delta^3 p$ are constrained by the condition

$$\Delta V \Delta^3 p \geq h^3. \tag{3.30}$$

Consider now an ideal electron gas of temperature T; the temperature determines the distribution of momenta according to equation (3.14). In particular, the *average* momentum (or velocity) is uniquely determined by T. Imagine now that the gas is compressed. The volume occupied by each particle, $\Delta V \propto \rho^{-1}$, decreases. As long as the temperature is sufficiently high (and, with it, the average velocity, or momentum), so that compression (decrease of ΔV) does not lead to the violation of Heisenberg's principle, we should be allowed to ignore quantum effects. Eventually, however, the density may become high (ΔV low) enough for the range of momenta dictated by the uncertainty principle to exceed the momentum corresponding to the gas temperature. Practically, this means that the electron pressure must be higher than that inferred from the temperature. In order to estimate the electron pressure under these conditions, we have to take account of another quantum mechanics principle, the *Pauli exclusion principle*, which postulates that no two electrons can occupy the same *quantum state*, that is, have the same momentum and the same spin. Since an electron can have two spin states (*up* and *down*), this means that each element of *phase space* – location and momentum space – can be occupied by two electrons at most. The pressure generated by electrons that are forced into higher momentum states as their density increases is called *degeneracy pressure*. A state of *complete degeneracy* is obtained when all the available momentum states are occupied up to a maximum momentum value. In this case $\Delta V \Delta^3 p$ is minimal and condition (3.30) becomes an equality. Such an ideal situation can only be achieved at zero temperature, but it constitutes a good approximation to states of high degeneracy and has the advantage of enabling a straightforward calculation of the pressure. Therefore, although the transition from a Maxwellian to a completely degenerate momentum distribution occurs gradually, we shall discuss only the extreme situations.

Applying the Heisenberg and Pauli principles to a completely degenerate isotropic electron gas yields the momentum distribution – the number of electrons with momenta in the interval $(p, p + dp)$ per unit volume:

$$n_e(p)\,dp = \frac{2}{\Delta V} = \frac{2}{h^3} 4\pi p^2\,dp, \qquad p \le p_0. \tag{3.31}$$

The maximal momentum p_0 can be obtained by integration, $n_e = \int_0^{p_0} n_e(p)\,dp$, and reversing the relation between n_e and p_0:

$$p_0 = \left(\frac{3h^3 n_e}{8\pi}\right)^{1/3}. \tag{3.32}$$

We may now use theorem (3.4), substituting equation (3.31), taking $v = p/m_e$, where m_e is the electron mass, and carrying the integral up to p_0, to obtain the degeneracy pressure of an electron gas

$$P_{e,\text{deg}} = \frac{8\pi}{15 m_e h^3} p_0^5 = \frac{h^2}{20 m_e}\left(\frac{3}{\pi}\right)^{2/3} \frac{1}{m_H^{5/3}} \left(\frac{\rho}{\mu_e}\right)^{5/3}, \tag{3.33}$$

where we have used relation (3.24) for μ_e. We note that the degeneracy pressure is inversely proportional to the particle (electron) mass. Hence, although the arguments presented for electrons could be equally applied to protons and neutrons, as they are nearly 2000 times more massive than electrons, quantum effects become important in their case under much more extreme conditions (much higher densities for a given temperature, and much lower temperatures for a given density), and may usually be ignored. We also note that, in spite of the high densities characteristic of degenerate matter, the particles may be still considered *free*, since the particle energy, of the order of $p_0^2/2m_e$, is still higher than the Coulomb energy ϵ_C.

> **Exercise 3.1:** Find the condition that the electron number density n_e must satisfy, for a degenerate electron gas to be considered perfect.

Inserting the numerical values of constants in equation (3.33), we obtain

$$P_{e,\text{deg}} = K_1'\left(\frac{\rho}{\mu_e}\right)^{5/3}, \tag{3.34}$$

where $K_1' = 1.00 \times 10^7 \frac{\text{N m}^{-2}}{(\text{kg m}^{-3})^{5/3}}$ [$1.00 \times 10^{13} \frac{\text{dyn cm}^{-2}}{(\text{g cm}^{-3})^{5/3}}$]. For a composition devoid of hydrogen (and not very rich in extremely heavy elements), $\mu_e \approx 2$, and hence the degeneracy pressure (3.33) is simply given by

$$P_{e,\text{deg}} = K_1 \rho^{5/3}, \tag{3.35}$$

where K_1 is a constant. This relation will be often used in future discussions.

If the electron density is increased further, the maximal momentum in a completely degenerate electron gas grows larger. Eventually, a density is reached such that the velocity p_0/m_e approaches the speed of light. The electrons now constitute a *relativistic* degenerate gas, for which the simple relation between momentum and velocity $p = mv$ no longer holds and has to be replaced by the relativistic kinematics relation. Here again, for simplicity, we shall only consider the extreme case in which the velocity is very close to c. Replacing v by c in the pressure integral (3.4), we obtain by the same procedure as before

$$P_{e,r-deg} = \frac{hc}{8} \left(\frac{3}{\pi}\right)^{1/3} \frac{1}{m_H^{4/3}} \left(\frac{\rho}{\mu_e}\right)^{4/3} \tag{3.36}$$

for the pressure exerted by a completely degenerate relativistic electron gas in the limit $v \to c$. The transition between relations (3.33) and (3.36) is a smooth function of v/c or $p/m_e c$, which we shall not address. Inserting the numerical values of constants in equation (3.36), we obtain

$$P_{e,r-deg} = K_2' \left(\frac{\rho}{\mu_e}\right)^{4/3}, \tag{3.37}$$

where $K_2' = 1.24 \times 10^{10} \frac{\text{N m}^{-2}}{(\text{kg m}^{-3})^{4/3}}$ [$1.24 \times 10^{15} \frac{\text{dyn cm}^{-2}}{(\text{g cm}^{-3})^{4/3}}$] and finally, for a fixed value of μ_e,

$$P_{e,r-deg} = K_2 \rho^{4/3}, \tag{3.38}$$

where K_2 is another constant.

We should keep in mind that relations (3.35) and (3.38) for the pressure were obtained on the assumption of vanishing temperature, and hence, naturally, the pressure is only a function of density (for a given composition). It is true, however, that even in the case of incomplete – or *partial* – degeneracy, the temperature plays a far smaller role than in the case of an ideal (nondegenerate) gas. Thus as a crude approximation, the degeneracy pressure may be considered as insensitive to temperature. The approximation is good provided kT is only a fraction of the kinetic energy of a particle with the highest momentum $p_0(n_e)$.

3.4 The radiation pressure

Radiation pressure is due to photons that transfer momentum to gas particles whenever they are absorbed or scattered. In thermodynamic equilibrium the photon distribution is isotropic and the number of photons with frequencies in the range

$(\nu, \nu + d\nu)$ is given by the Planck (blackbody distribution) function

$$n(\nu)\,d\nu = \frac{8\pi \nu^2}{c^3} \frac{d\nu}{e^{\frac{h\nu}{kT}} - 1}. \tag{3.39}$$

The pressure is then readily obtained from equation (3.4):

$$P_{\text{rad}} = \tfrac{1}{3} \int_0^\infty c \frac{h\nu}{c} n(\nu)\,d\nu = \tfrac{1}{3} a T^4, \tag{3.40}$$

where a is the radiation constant

$$a = \frac{8\pi^5 k^4}{15 c^3 h^3} = \frac{4\sigma}{c}. \tag{3.41}$$

Although the expression for radiation pressure was easily derived from the pressure integral, the concept deserves further (intuitive) explanation. Imagine a collimated beam of photons striking an atom. Each photon is absorbed, thereby exciting the atom, which consequently returns to its original state by emitting a photon. The direction of the emitted photon is random, the initial direction of the absorbed one having been "forgotten." Each such interaction involves an exchange of momentum. By absorbing the photon, the atom gains momentum in the direction of the photon beam. When it emits a photon, the atom recoils in the direction opposite to that of the emitted photon. After a long series of such interactions, the random changes of momentum due to emission cancel out and the net change in the atom's momentum is in the direction of the photon beam, as if material pressure has been exerted on it in that direction.

3.5 The internal energy of gas and radiation

The specific energy (energy per unit mass) of a perfect gas, which is due to the kinetic energy ϵ of the motion of the individual particles, is generally given by

$$u = \frac{1}{\rho} \int_0^\infty n(p)\,\epsilon(p)\,dp, \tag{3.42}$$

where the integral represents the energy density (energy per unit volume). For a classical gas, $\epsilon = p^2/2m_g$; for a relativistic gas

$$\epsilon = m_g c^2 \left[\left(1 + \frac{p^2}{m_g^2 c^2}\right)^{1/2} - 1 \right], \tag{3.43}$$

which tends to $p^2/2m_g$ in the limit $p \ll m_g c$. Performing the integral for a simple classical ideal gas, we obtain for the energy density the well-known result $\frac{3}{2}nkT$, which is equivalent to $\frac{3}{2}P$. The specific energy is therefore

$$u_{\text{gas}} = \frac{3}{2}\frac{P_{\text{gas}}}{\rho}. \tag{3.44}$$

For a classical completely degenerate electron gas we obtain the specific energy by integrating equation (3.42) up to the highest momentum p_0, and the result is identical with that obtained for the classical ideal gas – equation (3.44). For the relativistic completely degenerate case we obtain by the same procedure

$$u_{\text{gas}} = 3\frac{P_{\text{gas}}}{\rho}. \tag{3.45}$$

The energy density of radiation is given by

$$\int_0^\infty h\nu n(\nu)\, d\nu = aT^4, \tag{3.46}$$

where the integral is the same as in equation (3.40), the specific energy being

$$u_{\text{rad}} = \frac{aT^4}{\rho} = 3\frac{P_{\text{rad}}}{\rho}. \tag{3.47}$$

Exercise 3.2: Assuming a uniform value of β throughout the star and defining $U = \int (u_{\text{gas}} + u_{\text{rad}})dm$, show that the virial theorem (2.23) leads to

$$E = \frac{\beta}{2}\Omega = -\frac{\beta}{2-\beta}U$$

for a classical (nonrelativistic) gas. Note in particular the limits $\beta \to 1$ and $\beta \to 0$.

3.6 The adiabatic exponent

A special kind of thermodynamic processes, which will be of interest in later discussions, are those occurring in a system without exchange of heat with the environment. Such processes are called *adiabatic*. From the first law of thermodynamics (mentioned in Section 2.2) it follows that adiabatic processes satisfy the condition

$$du + Pd\left(\frac{1}{\rho}\right) = 0. \tag{3.48}$$

In the previous section we have seen – at least for simple systems – that the specific energy u is always proportional to P/ρ. We may therefore write

$$u = \phi \frac{P}{\rho}, \tag{3.49}$$

which, by differentiating and substituting in equation (3.48), leads to

$$\phi\, Pd\left(\frac{1}{\rho}\right) + \phi\frac{1}{\rho}dP + Pd\left(\frac{1}{\rho}\right) = (\phi+1)Pd\left(\frac{1}{\rho}\right) + \phi\frac{1}{\rho}dP = 0. \tag{3.50}$$

Accordingly, the dependence of the pressure on density is described by a power law

$$P \propto \rho^{\frac{\phi+1}{\phi}}. \tag{3.51}$$

The power $(d\ln P/d\ln\rho)$ is called the *adiabatic exponent*, denoted γ_a; the proportionality factor (to be denoted K_a) is determined by the properties of the system (it is a direct function of the *entropy*). In conclusion, adiabatic processes are characterized by the law

$$P = K_a \rho^{\gamma_a}. \tag{3.52}$$

It is easily seen that for the systems we have considered, the value of γ_a is 5/3 in the case of a nonrelativistic ideal gas or completely degenerate electron gas, and 4/3 in the case of a relativistic degenerate electron gas or of pure radiation. Intermediate values will obtain for mixtures, such as gas and radiation, and for nonextreme cases, such as a moderately relativistic degenerate electron gas.

So far we have considered gases of a fixed number of particles: either (almost) fully ionized, as in the deep stellar interior, or (almost) fully recombined, as in the outer layers of a cool stellar atmosphere. When ionization takes place and the number of particles changes with the other physical properties, the adiabatic exponent changes too. Since this will prove to be of particular importance to the stability of stars, it deserves some discussion. We shall only consider the very simple case of a singly ionized pure gas (rather than a mixture of gases), say, hydrogen. Hence we have to deal with three different types of particles: neutral atoms, whose number density we denote by n_0, ions of number density n_+, and free electrons of number density n_e (obviously, $n_e = n_+$). The pressure exerted

3.6 The adiabatic exponent

by the gas is proportional to $n_0 + n_+ + n_e$, while the mass density is proportional to $n_0 + n_+$. The *degree of ionization* is defined by

$$x = \frac{n_+}{n_0 + n_+}. \qquad (3.53)$$

The densities of ions and neutrals are related by Saha's equation (after Meghnad Saha, who derived it in 1920),

$$\frac{n_+ n_e}{n_0} = \frac{g}{h^3}(2\pi m_e kT)^{3/2} e^{-\chi/kT}, \qquad (3.54)$$

where g is a constant and χ is the ionization potential (the energy required to create an ion by removing an electron from an atom). In terms of the degree of ionization, we have

$$P = (1+x)(n_0 + n_+)kT = (1+x)\mathcal{R}\rho T, \qquad (3.55)$$

and Saha's equation becomes

$$\frac{x^2}{1-x^2} = \frac{g}{h^3} \frac{(2\pi m_e)^{3/2}(kT)^{5/2}}{P} e^{-\chi/kT}. \qquad (3.56)$$

In the case of a partially ionized gas, the specific energy has an additional term, $\chi n_+/\rho = \chi n_+/[(n_0 + n_+)m_H] = \chi x/m_H$, which is due to the available potential energy of ionization. Thus

$$u = \frac{3}{2}\frac{P}{\rho} + \frac{\chi}{m_H}x \qquad (3.57)$$

replaces equation (3.49). Using equations (3.55) and (3.56) to express the degree of ionization as a function of pressure and density $x = x(P, \rho)$, differentiating equation (3.57), and substituting into equation (3.48) yields

$$\frac{3}{2}\left(\frac{1}{\rho}\right)dP + \frac{3}{2}Pd\left(\frac{1}{\rho}\right) + \frac{\chi}{m_H}\frac{\partial x}{\partial P}dP + \frac{\chi}{m_H}\frac{\partial x}{\partial \rho}d\rho + Pd\left(\frac{1}{\rho}\right) = 0. \qquad (3.58)$$

Multiplying by ρ/P and assembling terms, we have

$$\left[\frac{3}{2} + \frac{\chi}{kT}\left(\frac{P}{1+x}\right)\left(\frac{\partial x}{\partial P}\right)_\rho\right]\frac{dP}{P} - \left[\frac{5}{2} - \frac{\chi}{kT}\left(\frac{\rho}{1+x}\right)\left(\frac{\partial x}{\partial \rho}\right)_P\right]\frac{d\rho}{\rho} = 0, \qquad (3.59)$$

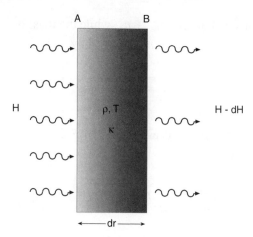

Figure 3.2 Radiation flux passing through a slab.

from which, after not inconsiderable manipulation, $\gamma_a(x)$ may be calculated:

$$\gamma_a(x) = \frac{5 + \left(\frac{5}{2} + \frac{\chi}{kT}\right)^2 x(1-x)}{3 + \left[\frac{3}{2} + \left(\frac{3}{2} + \frac{\chi}{kT}\right)^2\right] x(1-x)}. \qquad (3.60)$$

In the limit $x = 0$ or $x = 1$, we obtain $\gamma_a = 5/3$, as before; the minimum value is obtained for $x = 0.5$; for $\chi/kT = 10$, for example, it is $\gamma_a = 1.21$.

3.7 Radiative transfer

Consider a slab of thickness dr and density ρ between parallel surfaces A and B of unit area, as shown in Figure 3.2. A radiation flux H (energy per unit area per unit time) incident on A emerges at B after losing an amount dH, which has been absorbed by the slab (in reality, the slab may also emit radiation, which would have to be taken into account). Obviously, the amount of absorbed radiation should be proportional to the incident flux and to the amount of material, that is, to the density of absorbing (scattering) particles and to the length of the path travelled by the photons, and hence to their product $H\rho dr$. We may therefore write:

$$dH = -\kappa H \rho \, dr, \qquad (3.61)$$

where the minus sign indicates that the flux has been diminished, and κ is a constant, called the *opacity coefficient*, determined by the properties of the material the slab is made of, such as composition, density, and temperature. Integrating, we obtain

$$H = H_0 e^{-\kappa \rho r} \qquad (3.62)$$

3.7 Radiative transfer

for the radiation flux at a distance r from a source H_0. The characteristic absorption length, $(\kappa\rho)^{-1}$, may be regarded as the *mean free path* of a photon. The dimensionless quantity τ, defined by $d\tau \equiv -\kappa\rho\, dr$, is called *optical depth*; it is a measure of the transparency of a medium to radiation. An opaque medium has a large optical depth, which may be due to a large physical depth, a high opacity, or a high density, or to a combination of these factors. A transparent medium, which lets through most of the radiation crossing it, has a low optical depth. In a star, the concept of optical depth serves to define the photosphere. Being a gaseous sphere, a star does not have a well-defined surface; the stellar radius is, by definition, the radius of the surface where $T = T_{\text{eff}}$. To find this surface, we recall that the bulk of stellar radiation is emitted from the region lying above R, which is the photosphere; hence the optical depth of the photosphere, $\int_R^\infty \kappa\rho\, dr$, must be of the order of unity. This condition may be regarded as a definition of R, the exact value of the photospheric optical depth being determined by a detailed treatment of radiative transfer and its inherent assumptions and approximations.

> **Exercise 3.3:** Show that the equation of hydrostatic equilibrium may be written as
>
> $$\frac{dP}{d\tau} = \frac{g}{\kappa},$$
>
> where g is the local gravitational acceleration. (This form is useful in models of stellar atmospheres.)

As photons of different frequencies interact differently with matter, the opacity coefficient is also a function of the frequency of the radiation and the foregoing discussion applies strictly only to monochromatic radiation. It is, however, possible to define an average opacity, independent of wavelength (see Appendix 1). The most important interactions between stellar matter (of high temperature) and radiation are those involving electrons (rather than the much heavier nuclei). These are of several types.

(a) *Electron scattering* – the scattering of a photon by a free electron, the photon's energy remaining unchanged [known as Thomson scattering (after J. J. Thomson). In the less common, relativistic case, known as Compton scattering, the photon's energy does change.]
(b) *Free-free absorption* – the absorption of a photon by a *free* electron, which makes a transition to a higher energy state by briefly interacting with a nucleus or an ion. The inverse process, leading to the emission of a photon, is known as *bremsstrahlung*.
(c) *Bound-free absorption* – which is another name for photoionization – the removal of an electron from an atom (ion) caused by the absorption of a photon. The inverse process is radiative recombination.

(d) *Bound-bound absorption* – the excitation of an atom that is due to the transition of a (bound) electron to a higher energy state by the absorption of a photon. The atom is then de-excited either spontaneously or by collision with another particle, whereby a photon is emitted.

In the deep stellar interiors, where temperatures are very high, the first two processes are dominant, simply because there are very few bound electrons, the material being almost completely ionized. Furthermore, the energy of most photons in the Planck distribution is of the order of keV, whereas the separation energy of atomic levels is only a few tens eV. Hence most photons interacting with bound electrons would set them free. Thus bound-bound (and even bound-free) transitions have extremely low probabilities, interactions occurring predominantly between photons and *free* electrons.

Opacity coefficients may be measured or – for conditions typical of stellar interiors – calculated, taking into account all the possible interactions between different elements and photons of different frequencies. This is a tedious task that requires an enormous amount of calculation. When it has been performed, the results are usually approximated by relatively simple formulae in the form of power laws in density and temperature for a given composition:

$$\kappa = \kappa_0 \rho^a T^b. \tag{3.63}$$

The opacity resulting from electron scattering is temperature and density independent ($a = b = 0$); it is given by

$$\kappa_{es} = \frac{\kappa_{es,0}}{\mu_e} \approx \frac{1}{2}\kappa_{es,0}(1 + X), \tag{3.64}$$

where $\kappa_{es,0} = 0.04$ m^2 kg^{-1} (0.4 cm^2 g^{-1}). The opacity resulting from free-free absorption, first derived by Hendrik A. Kramers, is well approximated by a power law of the form of equation (3.63), with $a = 1$ and $b = -7/2$, known as the *Kramers opacity law*,

$$\kappa_{ff} = \frac{\kappa_{ff,0}}{\mu_e}\left\langle\frac{Z^2}{A}\right\rangle \rho T^{-7/2} \approx \frac{1}{2}\kappa_{ff,0}(1 + X)\left\langle\frac{Z^2}{A}\right\rangle \rho T^{-7/2}, \tag{3.65}$$

where complete ionization is assumed and $\langle\frac{Z^2}{A}\rangle = \sum_i X_i \frac{Z_i^2}{A_i}$. The constant is $\kappa_{ff,0} = 7.5 \times 10^{21}$ m^2 kg^{-1} (7.5×10^{22} cm^2 g^{-1}) with an accuracy of about 20%. Electron scattering and free-free opacities are both due to the free electrons; both coefficients, κ_{es} and κ_{ff}, are thus proportional to the electron number density and hence to μ_e^{-1} [see equation (3.24)]. Opacity coefficients for solar composition material are given, as an example, in Figure 3.3; note that above a few 10^4 K they may indeed be quite accurately represented by power laws.

3.7 Radiative transfer

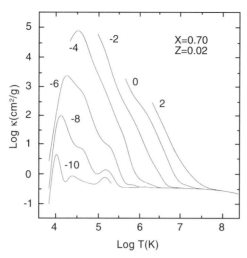

Figure 3.3 Opacity coefficients (in units of cm² g⁻¹) for a solar composition as a function of temperature for different density values; the numbers beside each curve are log ρ(g cm⁻³) [data from C. A. Iglesias & F. J. Rogers (1996), *Astrophys. J., 464*].

The average opacity of stellar material (of solar composition) is of the order of 0.1 m² kg⁻¹ (1 cm² g⁻¹), and since the average density is of the order of 1000 kg m⁻³ (1 g cm⁻³), the mean free path of photons in the interior of a star is about 0.01 m (1 cm). The temperature drop over such a radial distance within a star is about 0.001 K (estimated as $\bar{T}/R$). This is why the radiation in stellar interiors is so close to that of a blackbody. But blackbody radiation is also isotropic; what, then, is the meaning of a radiation flux from the interior of the star to the surface [the function F in equation (2.6)]? It turns out that the minute deviation from isotropy is sufficient for the transfer of energy that results in the stellar luminosity.

To calculate the radiative flux we shall adopt a simple approach due to Eddington. The absorption of radiation energy by the slab just considered also involves a corresponding amount of momentum: the momentum absorbed by the slab per unit time is $|dH|/c$. The rate of increase of the momentum must be equal to the net force applied to the slab by the radiation field (Newton's second law). This force is simply the difference of the radiation pressures exerted on the surfaces A, say at r, and B, at $r+dr$ (see Figure 3.2): $P_{\text{rad}}(r) - P_{\text{rad}}(r+dr) = -(dP_{\text{rad}}/dr)\,dr$. Consequently,

$$\frac{H\kappa\rho}{c} = -\frac{dP_{\text{rad}}}{dr}, \qquad (3.66)$$

and since the radiation may be assumed to be blackbody radiation, the pressure is given by equation (3.40) and

$$H = -\frac{4acT^3}{3\kappa\rho}\frac{dT}{dr}. \qquad (3.67)$$

A rigorous derivation of this equation, leading to the correct evaluation of the average opacity, is given in Appendix 1. To obtain the total flux F crossing a spherical surface of radius r, we multiply H by the surface area $4\pi r^2$:

$$F = -4\pi r^2 \frac{4acT^3}{3\kappa\rho} \frac{dT}{dr}. \tag{3.68}$$

We may invert this relation to obtain the temperature gradient in terms of the flux:

$$\frac{dT}{dr} = -\frac{3}{4ac} \frac{\kappa\rho}{T^3} \frac{F}{4\pi r^2}, \tag{3.69}$$

or, using m as the independent space variable,

$$\frac{dT}{dm} = -\frac{3}{4ac} \frac{\kappa}{T^3} \frac{F}{(4\pi r^2)^2}. \tag{3.70}$$

We have now gathered sufficient information on the physics of stellar interiors to allow us to pursue the investigation of stellar evolution.

4

Nuclear processes that take place in stars

The evolution – continuous change – of stars is due to their sustained emission of radiation originating from an internal source. The energy source that supplies the luminosity of stars during most of their lifetimes is nuclear fusion, which turns a small fraction of the rest mass into energy. Although this was only realized at the beginning of the twentieth century, with Einstein's formula $E = mc^2$, the *concept* of conversion of matter into light dates back to Newton, at the beginning of the eighteenth century.

> "The changing of bodies into light, and light into bodies, is very conformable to the course of Nature, which seems delighted with transmutations."
> Isaac Newton: Opticks, 1704

The formalism by which nuclear reactions are incorporated into the stellar evolution theory was given in Section 2.6. The purpose of the present chapter is to examine in more detail the nuclear processes that are bound to take place in stars and the energy each of them can supply.

4.1 The binding energy of the atomic nucleus

The energy released or absorbed in a nuclear reaction being a fraction of the rest-mass energy of the particles involved, mass is not strictly conserved: the total mass of the products differs slightly from the total mass of the reactants, the difference depending on the binding energies of the interacting nuclei. As we have seen in Section 2.6, the general description of a nuclear reaction is

$$I(\mathcal{A}_i, \mathcal{Z}_i) + J(\mathcal{A}_j, \mathcal{Z}_j) \rightleftharpoons K(\mathcal{A}_k, \mathcal{Z}_k) + L(\mathcal{A}_l, \mathcal{Z}_l). \tag{2.48}$$

Denoting by Q_{ijk} the amount of energy released in this reaction, and by $\mathcal{M}_i$ the mass of a nucleus of type I, we have

$$Q_{ijk} = (\mathcal{M}_i + \mathcal{M}_j - \mathcal{M}_k - \mathcal{M}_l)c^2, \tag{4.1}$$

neglecting the small masses of possible light particles that may be involved. Using the mass unit m_H, we may write equation (4.1) as

$$Q_{ijk} = [(\mathcal{M}_i - \mathcal{A}_i m_H) + (\mathcal{M}_j - \mathcal{A}_j m_H) - (\mathcal{M}_k - \mathcal{A}_k m_H) \\ - (\mathcal{M}_l - \mathcal{A}_l m_H)]c^2 + (\mathcal{A}_i + \mathcal{A}_j - \mathcal{A}_k - \mathcal{A}_l)m_H c^2, \tag{4.2}$$

where the second term on the right-hand side vanishes, by conservation of baryon number, equation (2.49). The difference

$$\Delta\mathcal{M}(I) \equiv (\mathcal{M}_i - \mathcal{A}_i m_H)c^2 \tag{4.3}$$

(whether positive or negative) is called *mass excess*, despite its being a measure of energy. Mass excesses are listed in tables of nuclear data in units of MeV (1 MeV $= 10^6$ eV $= 1.6021772 \times 10^{-13}$ J). We note that mass-excess values depend on the atomic mass unit employed, but Q_{ijk}, which involves differences of mass excesses, is independent of the (arbitrary) mass unit.

We may now calculate the total rate of energy release at a given point in a star: since the number of reactions of type (2.48) that occur per unit volume per unit time is $n_i n_j R_{ijk}$ (if $I = J$, $n_i n_j$ should be replaced by $\frac{1}{2}n_i^2$), the energy released by such reactions per unit volume per unit time is $n_i n_j R_{ijk} Q_{ijk}$. Summing over all nuclear reactions that can occur at that point, and dividing by ρ to obtain the rate of energy released per unit mass, we have

$$q = \frac{\rho}{m_H^2} \sum_{ijk} \frac{1}{1+\delta_{ij}} \frac{X_i}{\mathcal{A}_i} \frac{X_j}{\mathcal{A}_j} R_{ijk} Q_{ijk} \tag{4.4}$$

for the term that appears on the right-hand side of the energy equation (2.6). In reality, the available energy (to be turned into heat) may be less. If neutrinos are produced by the nuclear reactions (or by other processes), their energy is lost to the star, which is transparent to neutrinos. These "particles" leave the star without undergoing collisions and sharing their energy with the medium. Therefore, the net rate of energy release is $q_{\text{nuc}} - q_\nu$, where q_{nuc} is given by equation (4.4) and q_ν is the neutrino energy lost per unit mass per unit time.

Neutrinos are not only produced in nuclear reactions involving electrons and positrons, but also in interactions similar to those of an electron with the radiation field (photons), in which a change in the electron's momentum occurs.

Simply, the emerging photon is sometimes – usually extremely seldom – replaced by a neutrino-antineutrino pair. Thus *photoneutrinos* are produced when a photon is scattered by an electron, replacing the outgoing photon. The annihilation of an electron-positron pair (that we shall discuss in Section 4.9), which normally results in the creation of two photons, may produce a neutrino-antineutrino pair instead, with a probability of about 10^{-19}. Bremsstrahlung photons, emitted when an electron is decelerated by the Coulomb field of a nucleus or an ion (see Section 3.7), may also be replaced by a neutrino-antineutrino pair. Finally, a photon may itself decay into a neutrino-antineutrino pair, when the radiation field is affected by the electromagnetic field of the stellar plasma. The role of the electron is replaced in this case by a virtual particle called *plasmon* – essentially, a quantized plasma wave. All these processes become important either at very high densities or at very high temperatures (or both). Then, because of the transparency of stellar material to neutrinos (the mean free path of stellar neutrinos is about $10^9 R_\odot$!), they may cause efficient local cooling.

The energy released in a nuclear reaction Q_{ijk} is a measure of the difference between the binding energies of the reactants and the products. The total binding energy of a nucleus is, of course, a function of the number of nucleons; but even the binding energy *per nucleon* differs from element to element, or among isotopes of the same chemical element (nuclei with the same $\mathcal{Z}$, but different $\mathcal{A}$). This implies that some nuclear structures are more *stable* than others. There are also *unstable* nuclei that spontaneously decay by emitting light particles such as electrons (β^- decay) or positrons (β^+ decay) – these are called *radioactive isotopes*; and there are *excited* nuclei, in a high-energy state, that emit energetic photons, thereby becoming more strongly bound. The nuclear-shell model, similar in many respects to the electron-shell model of the atom, explains these properties.

For our purposes, the important result is the variation of binding energy per nucleon with baryon number $\mathcal{A}$, shown in Figure 4.1, relative to the free proton – the hydrogen nucleus. The general trend is an increase of the binding energy per nucleon with atomic mass up to iron ($\mathcal{A} = 56$) and a slow monotonic decline beyond iron. The steep rise of the binding energy from hydrogen, through deuterium and ^{3}He, to ^{4}He implies that fusion of hydrogen into helium should release a large amount of energy per nucleon (unit mass), considerably larger than that released in, say, fusion of helium into carbon. Energy may be gained by *fusion* of light nuclei into heavier ones up to iron and, to a lesser extent, by breakup, or *fission* of heavy nuclei into lighter ones down to iron. In a different context, we recognize the first process as the basic mechanism of the H-bomb, and the second as the mechanism of the A-bomb. Another important fact related to the binding energy of atomic nuclei is that there are no stable configurations for $\mathcal{A} = 5$ and for $\mathcal{A} = 8$; ^{4}He is more tightly bound than its immediate neighbours.

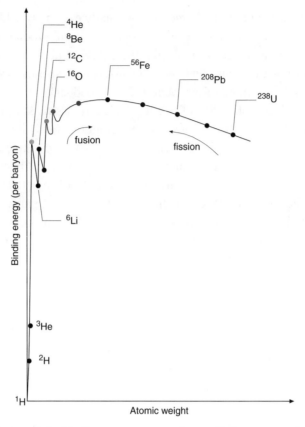

Figure 4.1 Variation of the binding energy per nucleon with baryon number.

4.2 Nuclear reaction rates

In Section 2.6 we have seen that the rate of a nuclear reaction is essentially the product of the cross-sectional area of a (target) nucleus and the relative velocity of the interacting gas particles. For the latter, we may simply assume a Maxwellian velocity distribution [equation (3.14)]: this means that the probability of the velocity of a particle of mass m_g being within an interval dv around a velocity v would be proportional to $\exp(-m_g v^2/2kT)$, decreasing with increasing v. Since, in reality, the target nuclei are not at rest (as assumed, for simplicity, in Section 2.6) and as v is the relative velocity of the interacting particles I and J, m_g is their reduced mass [$m_{g,i} m_{g,j}/(m_{g,i}+m_{g,j})$]. The term ς in the product presents a more difficult problem: in order to induce a nuclear reaction, nuclei have to come within a distance comparable to the range of the strong force. Since they are positively charged, to do so they must overcome the Coulomb repulsive force, which tends to separate them. This force imposes an effective barrier at a separation distance d, where the kinetic energy of the particles equals the electric potential

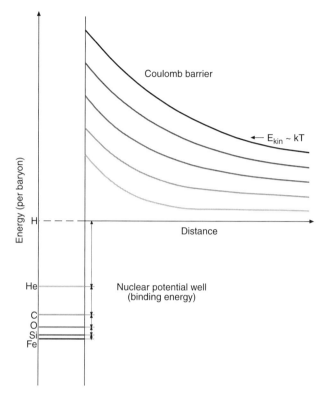

Figure 4.2 Schematic representation of the Coulomb barrier – the repulsive potential encountered by a nucleus in motion relative to another – and the short-range negative potential well that is due to the nuclear force. The height of the barrier and the depth of the well depend on the nuclear charge (atomic number).

energy,

$$d = \frac{1}{4\pi\varepsilon_0} \frac{Z_i Z_j e^2}{\frac{1}{2} m_g v^2}. \tag{4.5}$$

For average stellar temperatures (as derived in Section 2.4), the thermal velocities are such that the Coulomb barrier is set at a distance which is almost three orders of magnitude *larger* than the typical range of the strong nuclear force! This is illustrated schematically in Figure 4.2. In other words, the kinetic (thermal) energy of the gas in stellar interiors is of the order of keV, while the height of the Coulomb barrier at nuclear distances is of the order of MeV.

We can now understand why, during the first quarter of the twentieth century, it was thought impossible for such interactions to take place in stars: simply, stars appeared not to be sufficiently hot. The solution to this puzzle was provided by quantum mechanics. A rigorous explanation is beyond the scope of this text; suffice it to say that, according to quantum mechanics, there is a finite (nonvanishing) probability for a particle to penetrate the Coulomb barrier, as if a "tunnel" existed to carry it through. This quantum effect, discovered by George Gamow in 1928 in connection with radioactivity, is indeed called "tunnelling." It was applied to energy generation in stars by Robert Atkinson and Fritz Houtermans in 1929, soon after its discovery.

The penetration probability, as calculated by Gamow, and with it the nuclear cross section, is proportional to $\exp(-\pi Z_i Z_j e^2/\varepsilon_0 h v)$, thus increasing with v. In conclusion, the product $\exp(-\pi Z_i Z_j e^2/\varepsilon_0 h v) \exp(-m_g v^2/2kT)$, where the first exponent increases and the second decreases with increasing v, has a maximum known as the *Gamow peak*. To calculate the reaction rate, we would have to integrate the product over all velocity values. It can be shown that the value of the integral, and with it the reaction rate, is proportional to the maximum of the product, which occurs for

$$v = (\pi Z_i Z_j e^2 kT/\varepsilon_0 h m_g)^{1/3}. \tag{4.6}$$

Hence the reaction rate

$$\varsigma v \propto (kT)^{-2/3} \exp\left[-\frac{3}{2}\left(\frac{\pi Z_i Z_j e^2}{\varepsilon_0 h}\right)^{2/3}\left(\frac{m_g}{kT}\right)^{1/3}\right]$$

increases with increasing temperature and decreases with increasing charges of the interacting particles. Fusion of heavier and heavier nuclei would therefore require higher and higher temperatures. Reactions of a special type, called *resonant reactions*, interfere with this monotonic trend. They occur when the energy of the interacting particles corresponds to an energy level of the compound nucleus $(I+J)$, which is formed for a very brief period of time, before decaying into the reaction products K and L. In this case the reaction cross section has a very sharp peak at the resonant energy, several orders of magnitude higher than the cross sections at neighbouring energies.

The typical timescale of a nuclear reaction is inversely proportional to the reaction rate. For example, the characteristic time of destruction of type I nuclei by collisions with type J nuclei, leading to reactions of the form (2.48), would be given by

$$\tau_i = (n_j R_{ijk})^{-1}.$$

The extremely high sensitivity of nuclear reaction rates to temperature leads to the concept of "ignition" of a nuclear fuel: each reaction (or nuclear process) has a typical narrow temperature range over which its rate increases by orders of magnitudes, from negligible values to very significant ones. Around this range, the temperature dependence of the reaction rate may be well approximated by a power law (with a high power) and an *ignition* or *threshold* temperature may be defined. Hence, by equation (4.4), we should characteristically have $q \propto \rho T^n$, or

$$q = q_0 \rho T^n. \tag{4.7}$$

The process of creation of new nuclear species by fusion reactions is called *nucleosynthesis*. And since the kinetic energy of particles is that of their *thermal* motion, the reactions between them are called *thermonuclear*, as mentioned in Section 2.6. The simple chain of arguments presented here may be misleading; nuclear reaction rates involve quite complicated calculations, taking into account the particular structure (energy states) of the interacting nuclei. A detailed account of the physics of nuclear reactions may be found in Donald Clayton's classic book, *Principles of Stellar Evolution and Nucleosynthesis*, first published in 1968.

4.3 Hydrogen burning I: the p–p chain

The most abundant element in newly born stars is hydrogen, with $\mathcal{Z} = 1$. Fusion of hydrogen into the next element, helium, with $\mathcal{Z} = 2$, would require an encounter of three or four protons – hydrogen nuclei – within a distance of the order of fermis. The probability of such a multiple encounter is vanishingly small. Thus the process by which hydrogen is, eventually, turned into helium does not happen at once but gradually, through a chain of reactions, each involving the close encounter of only two particles. The first link of this chain should obviously be fusion of two protons (by the nuclear force – the *strong* interaction). The resulting particle would be, however, unstable and it would immediately disintegrate back into two separate protons. The way out of this impasse was found by Hans Bethe in 1939: during the close encounter of two protons, the *weak* interaction may convert one proton into a neutron, thus forming a heavier, stable isotope of hydrogen, deuterium:

$$p + p \to {}^2\text{D} + e^+ + \nu.$$

We note that all three conservation laws are obeyed – baryon number, lepton number, and charge. Then deuterium captures a proton to form the lighter helium isotope, ${}^3\text{He}$:

$$^2\text{D} + p \to {}^3\text{He} + \gamma,$$

where γ indicates the emission of an energetic photon, which will soon be absorbed, and whose energy will be shared by neighbouring particles. The chain now ramifies: one branch following the encounter of two ${}^3\text{He}$ isotopes, and the other, the encounter of a ${}^3\text{He}$ isotope with a ${}^4\text{He}$ one;

$$^3\text{He} + {}^3\text{He} \to {}^4\text{He} + 2p,$$

4 Nuclear processes that take place in stars

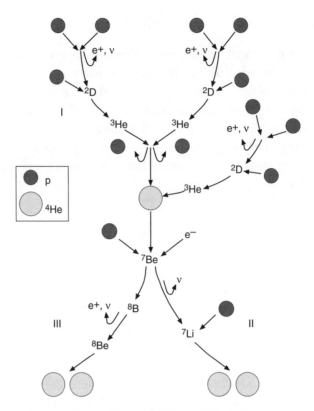

Figure 4.3 The nuclear reactions of the p–p I, II, and III chains.

or

$$^3\text{He} + {}^4\text{He} \rightarrow {}^7\text{Be} + \gamma.$$

The first branch marks the end of a chain – called the *p–p* I chain – that turns six protons into a ^{4}He nucleus (also known as an α particle), returning two protons, as illustrated in Figure 4.3. The second branch ramifies again, defining the *p–p* II and the *p–p* III chains shown in Figure 4.3. The *p–p* II chain proceeds with the capture of an electron by the beryllium nucleus, accompanied by the emission of a neutrino:

$$^7\text{Be} + e^- \rightarrow {}^7\text{Li} + \nu,$$

and the subsequent capture of another proton, to form *two* ^{4}He nuclei:

$$^7\text{Li} + p \rightarrow 2\,{}^4\text{He}.$$

The *p–p* III chain results from the capture by ^{7}Be of a proton, instead of an electron:

$$^7\text{Be} + p \to {}^8\text{B} + \gamma.$$

The radioactive boron isotope ^{8}B decays into ^{8}Be, which is highly unstable and immediately breaks into two ^{4}He nuclei:

$$^8\text{B} \to {}^8\text{Be} + e^+ + \nu$$

$$^8\text{Be} \to 2\,{}^4\text{He}.$$

This completes the p–p chain, whose three branches operate simultaneously. The relative importance of these chains, that is, the *branching ratios*, depend upon the conditions of hydrogen burning: temperature, density, and abundances of the elements involved. For example, for $X = Y$, the transition from p–p I to p–p II occurs gradually between temperatures of 1.3×10^7 K and 2×10^7 K; above 3×10^7 K the p–p III chain dominates. However, at such high temperatures, a different hydrogen burning process may favourably compete with the p–p chains, as we shall see shortly.

The energy released in the formation of an α particle by fusion of four protons is essentially given by the difference of the mass excesses of four protons and one α particle,

$$Q_{p-p} = 4\Delta\mathcal{M}(^1\text{H}) - \Delta\mathcal{M}(^4\text{He}) = 26.73 \text{ MeV},$$

according to the atomic mass table. Since any reaction chain that accomplishes this task must also turn two protons into neutrons, two neutrinos are emitted, which carry energy away from the reaction site. (In fact, it is these neutrinos that bear direct testimony to the occurrence of nuclear reactions in the interiors of stars, which would be otherwise unobservable. We shall return to this point in Section 8.3, when we discuss solar neutrinos.) The amounts of energy carried by the neutrinos vary for the different reaction chains: from 0.26 MeV for the creation of deuterium to 7.2 MeV for the boron decay. Since the p–p III chain, which includes the boron decay, has a small probability (branching ratio), 26 MeV are liberated on the average for each helium nucleus assembled, which, translated into energy per unit mass, yields 6×10^{14} J kg^{-1} (6×10^{18} erg g^{-1}).

Finally, the *rate* of energy release is determined by the *slowest* reaction in the chain, which is the first one, with a typical timescale of almost 10^{10} yr. It may be approximated by a power law in temperature with an exponent ranging from less than 4 and up to $\sim$6. Roughly, we may assume

$$q_{p-p} \propto \rho T^4. \tag{4.8}$$

Not only does the p–p chain require the lowest temperature among fusion processes, but it also exhibits the weakest temperature sensitivity.

4.4 Hydrogen burning II: the CNO bi-cycle

We have seen in Chapter 1 that a small percentage of the initial composition of any star consists of carbon, nitrogen, and oxygen (CNO) nuclei. These nuclei may induce a chain of reactions that transform hydrogen into helium, in which they themselves act similarly to catalysts in chemical reactions: they are destroyed and reformed in a cyclic process. The process, which is accordingly named the *CNO cycle*, was suggested by Bethe and, independently, by Carl-Friedrich von Weizsäcker, in 1938. The reactions involved are shown schematically in Figure 4.4. We note that here, too, as in the case of the *p–p* chain, the process may ramify, with the two different branches forming a *bi-cycle*. Each of the two closed chains that form the CNO bi-cycle involve six reactions resulting in the production of one ^{4}He nucleus: four proton captures and two β^+ decays accompanied by the emission of neutrinos. They are listed below, in parallel.

$$
\begin{array}{ll}
^{12}\text{C} + {}^{1}\text{H} \rightarrow {}^{13}\text{N} + \gamma & \qquad {}^{14}\text{N} + {}^{1}\text{H} \rightarrow {}^{15}\text{O} + \gamma \\
^{13}\text{N} \rightarrow {}^{13}\text{C} + e^+ + \nu & \qquad {}^{15}\text{O} \rightarrow {}^{15}\text{N} + e^+ + \nu \\
^{13}\text{C} + {}^{1}\text{H} \rightarrow {}^{14}\text{N} + \gamma & \qquad {}^{15}\text{N} + {}^{1}\text{H} \rightarrow {}^{16}\text{O} + \gamma \\
^{14}\text{N} + {}^{1}\text{H} \rightarrow {}^{15}\text{O} + \gamma & \qquad {}^{16}\text{O} + {}^{1}\text{H} \rightarrow {}^{17}\text{F} + \gamma \\
^{15}\text{O} \rightarrow {}^{15}\text{N} + e^+ + \nu & \qquad {}^{17}\text{F} \rightarrow {}^{17}\text{O} + e^+ + \nu \\
^{15}\text{N} + {}^{1}\text{H} \rightarrow {}^{12}\text{C} + {}^{4}\text{He} & \qquad {}^{17}\text{O} + {}^{1}\text{H} \rightarrow {}^{14}\text{N} + {}^{4}\text{He}
\end{array}
$$

We note that the number (total abundance) of CNO (and F) nuclei taking part in the process is constant in time; the relative abundances of the species depend upon the conditions of burning, mainly the prevailing temperature. The burning *rate* – as in any chain of reactions – is determined by the *slowest* reaction in the chain. In this context, it is important to note that, while β decays are independent of external conditions, capture reactions are extremely sensitive to temperature. Hence a very wide range of burning rates is to be expected, but only as long as capture reactions proceed more slowly than β decays. At the extremely high temperatures for which the situation is reversed, β decays would act as a bottleneck to the nuclear reaction sequence, regardless of temperature.

The energy released in the formation of a ^{4}He nucleus by the CNO cycle is ~25 MeV, after subtracting the energy carried away by the neutrinos. The temperature dependence of the energy generation rate q may be roughly approximated by a steep power law:

$$q_{\text{CNO}} \propto \rho T^{16}. \tag{4.9}$$

Thus, both processes of hydrogen burning – the main source of stellar energy – were brought to light at about the same time, and Bethe played a crucial role in both. Many years later, in 1967, he was awarded the Nobel Prize for Physics for his contribution to the understanding of energy production in stars.

4.5 Helium burning: the triple-α reaction

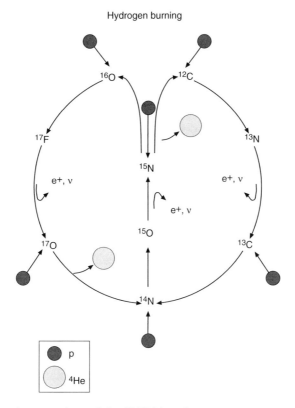

Figure 4.4 The nuclear reactions of the CNO bi-cycle.

4.5 Helium burning: the triple-α reaction

As in the case of hydrogen burning, the simplest and most obvious nuclear reaction in a helium gas should be fusion of two helium nuclei (α particles). But we have seen in Section 4.1 that there exists *no* stable nuclear configuration with $A = 8$ (regardless of $\mathcal{Z}$). Two helium nuclei may be fused into a beryllium isotope

$$^4\text{He} + {}^4\text{He} \rightarrow {}^8\text{Be},$$

but the ^{8}Be lifetime is only 2.6×10^{-16} of a second! The solution to this new problem was provided by Edwin Salpeter in 1952. Short as the ^{8}Be lifetime may seem, it is nevertheless *longer* than the mean collision (scattering) time of α particles at temperatures of the order of 10^8 K. Therefore, even at the seemingly negligible ^{8}Be abundance of one in 10^9 particles, there is a nonvanishing probability that an α particle will collide with a ^{8}Be nucleus before it decays, to produce carbon:

$$^8\text{Be} + {}^4\text{He} \rightarrow {}^{12}\text{C}.$$

Fred Hoyle realised shortly afterwards that the small probability of an α capture by a beryllium nucleus would be greatly enhanced if the carbon nucleus had an energy level close to the combined energies of the reacting ^{8}Be and ^{4}He nuclei. The reaction would then be a relatively fast *resonant* reaction. Remarkably, such a resonant energy level of ^{12}C was subsequently found experimentally in the Kellogg Radiation Laboratory at the California Institute of Technology.

Thus helium burning proceeds in a two-stage reaction that leads to the fusion of three helium nuclei into ^{12}C; hence the name of this reaction: *triple-α* (or 3α). The energy released in such a reaction is easily calculated:

$$Q_{3\alpha} = 3\Delta\mathcal{M}(^4\text{He}) - \Delta\mathcal{M}(^{12}\text{C}) = 7.275 \text{ MeV},$$

and, correspondingly, the energy generated per unit mass is 5.8×10^{13} J kg^{-1} (5.8×10^{17} erg g^{-1}). This is about one tenth of the energy generated by fusion of hydrogen into helium! The rate of this process is determined by the second reaction in the chain. It is thus proportional to the ^{8}Be abundance, which itself varies as the square of the helium abundance. Consequently, the energy generation rate depends on the *square* of the density. Its temperature sensitivity is quite astounding:

$$q_{3\alpha} \propto \rho^2 T^{40}. \tag{4.10}$$

When a sufficient number of carbon nuclei have accumulated by 3α reactions, it seems reasonable that α captures by these nuclei, and possibly by their products, could lead to the formation of heavier and heavier particles. In reality, it turns out that the increasing Coulomb barrier renders the probability of such captures very low compared with that of the 3α reaction, at least until the helium abundance becomes small. Hence the only significant α capture reaction that takes place is

$$^{12}\text{C} + {}^4\text{He} \rightarrow {}^{16}\text{O}.$$

The energy released by this reaction is 7.162 MeV, amounting to 4.3×10^{13} J kg^{-1}.

Note: It was the competition between the ^{12}C + ^{4}He and the ^{8}Be + ^{4}He reactions that led Hoyle to the prediction of the resonant energy level in the carbon nucleus. Already in 1946, with remarkable foresight, Hoyle had postulated that *all* nuclei (not only helium) build up from lighter nuclei by fusion reactions that take place in the interior of stars. Pursuing this idea, he considered the synthesis of elements from carbon to nickel, in 1953–1954, with the Salpeter process as starting point. He then showed that the observed cosmic abundance ratios He:C:O could be made to fit the yields calculated for the above reactions, *if* the ^{8}Be + ^{4}He reaction had a resonance corresponding to a level at ∼7.7 MeV in the ^{12}C nucleus. Otherwise, the inferred cosmic carbon abundance would be too low. Eager to test this prediction, Hoyle even collaborated in the first attempts to detect such a level experimentally.

4.6 Carbon and oxygen burning

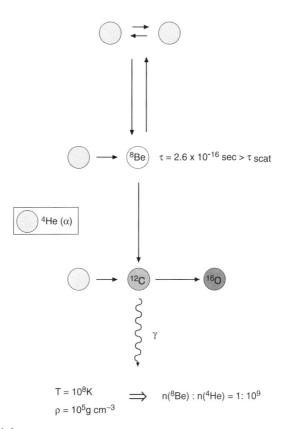

Figure 4.5 The triple-α process.

To summarize, the products of helium burning are carbon and oxygen, in relative abundances which depend on temperature. The process is shown schematically in Figure 4.5.

> **Exercise 4.1:** Calculate the energy generated per unit mass, if helium burning produces equal amounts (mass fractions) of carbon and oxygen.

4.6 Carbon and oxygen burning

Carbon burning – fusion of two carbon nuclei – requires temperatures above 5×10^8 K, and oxygen burning, having to overcome a still higher Coulomb barrier, occurs only at temperatures in excess of 10^9 K. Interactions of carbon and oxygen nuclei need not be considered, for at the intermediate temperature required by the intermediate Coulomb barrier, carbon nuclei are quickly exhausted by interacting with themselves.

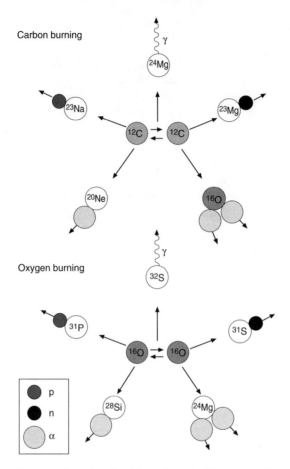

Figure 4.6 The nuclear reactions involved in carbon and in oxygen burning.

The processes of carbon and of oxygen burning are very similar: in both cases a compound nucleus is produced, at an excited energy level, and it subsequently decays. Several decay options are open, with different, temperature-dependent probabilities (branching ratios).

$$\begin{aligned} ^{12}\text{C} + ^{12}\text{C} &\rightarrow {}^{24}\text{Mg} + \gamma & ^{16}\text{O} + ^{16}\text{O} &\rightarrow {}^{32}\text{S} + \gamma \\ &\rightarrow {}^{23}\text{Mg} + n & &\rightarrow {}^{31}\text{S} + n \\ &\rightarrow {}^{23}\text{Na} + p & &\rightarrow {}^{31}\text{P} + p \\ &\rightarrow {}^{20}\text{Ne} + \alpha & &\rightarrow {}^{28}\text{Si} + \alpha \\ &\rightarrow {}^{16}\text{O} + 2\alpha & &\rightarrow {}^{24}\text{Mg} + 2\alpha \end{aligned}$$

The possible reaction channels are shown schematically in Figure 4.6.

On the average, 13 MeV are released for each $^{12}\text{C} + ^{12}\text{C}$ reaction and about 16 MeV for each $^{16}\text{O} + ^{16}\text{O}$ reaction, amounting to $\sim 5.2 \times 10^{13}$ J kg^{-1} and $\sim 4.8 \times 10^{13}$ J kg^{-1}, respectively. These reactions entail production of light particles, such as protons and helium nuclei, which are immediately captured by the heavy nuclei present, because of the relatively low Coulomb barriers. Thus many different isotopes are created by secondary reactions, besides those primarily produced by fusion of carbon or oxygen. The major nucleus formed by oxygen burning is silicon (^{28}Si), although other elements are also significantly abundant.

4.7 Silicon burning: nuclear statistical equilibrium

In principle, we may now assume by analogy that two silicon nuclei could fuse to create iron, the most stable element – the end product of the nuclear fusion chain. In reality, however, the Coulomb barrier has become prohibitively large. At temperatures above the oxygen burning range, but way below those that would be required for silicon fusion, another type of nuclear process takes place. It involves the interaction of massive particles with energetic photons, which are capable of disintegrating nuclei, much as less energetic photons are capable of breaking up atoms by tearing electrons away. The process, called *photodisintegration*, is similar in many respects to *photoionization* of atoms, except that the binding force is nuclear, instead of electric, and the emitted particles are light nuclei, instead of electrons. As in the case of ionization, reactions can proceed both ways and equilibrium may be achieved, with relative abundances depending on the prevailing physical conditions. The reaction $^{16}O + \alpha \rightleftharpoons {}^{20}Ne + \gamma$, for example, produces neon at temperatures around 10^9 K, but it reverses direction above 1.5×10^9 K. The energy absorbed in the inverse reaction (photodisintegration) is supplied by the radiation field.

Silicon disintegration occurs around 3×10^9 K; the light particles emitted are recaptured by other silicon nuclei, building up an entire network of nuclear reactions, with light particles exchanged between heavy nuclei. Although the nuclear reactions tend to equilibrium, where direct and inverse reactions occur at (almost) the same rate, the resulting state of *nuclear statistical equilibrium* is not perfect: a leakage occurs toward the stable iron group nuclei (Fe, Co, Ni), which resist photodisintegration until the temperature reaches $\sim 7 \times 10^9$ K.

The major nuclear burning processes that we have encountered and their main characteristics are summarized in Table 4.1. Their common feature is the release of energy upon consumption of nuclear fuel. The amounts and the rates of energy release vary, however, enormously. But nuclear processes that *absorb* energy (from the radiation field) are also possible under conditions expected to occur in stellar interiors. Their consequences may range from mild to catastrophic, depending on the amount of absorbed energy and, especially, on the rate of energy absorption. Such are the processes discussed in the following sections.

> **Exercise 4.2:** Estimate the minimal stellar mass required for the central ignition of the different nuclear fuels, according to the threshold temperatures of Table 4.1, by assuming (a) a density profile as in Exercise 1.3; (b) solar composition; (c) nondegeneracy.

Table 4.1 Major nuclear burning processes

Nuclear Fuel	Process	$T_{threshold}$ 10^6 K	Products	Energy per Nucleon (MeV)
H	p–p	~4	He	6.55
H	CNO	15	He	6.25
He	3α	100	C, O	0.61
C	C + C	600	O, Ne, Na, Mg	0.54
O	O + O	1000	Mg, S, P, Si	~0.3
Si	Nuc. eq.	3000	Co, Fe, Ni	<0.18

4.8 Creation of heavy elements: the s- and r-processes

So far we have considered charged particle interactions, their rates being controlled by the height of the Coulomb barrier, and interactions of nuclei with photons, which become efficient at high temperatures. Another type of interaction becomes possible in the presence of free neutrons, which are produced during carbon, oxygen, and silicon burning. *Neutron capture* by relatively heavy nuclei is not limited by the Coulomb barrier and can therefore proceed at relatively low temperatures. The only obstacle in the way of neutron capture reactions is the scarcity of neutrons.

Suppose a sufficient number density of neutrons is available. A chain of reactions would then be triggered, with nuclei capturing more and more neutrons, thus creating heavier and heavier isotopes of the same element:

$$I(A, Z) + n \rightarrow I_1(A + 1, Z),$$
$$I_1(A + 1, Z) + n \rightarrow I_2(A + 2, Z),$$
$$I_2(A + 2, Z) + n \rightarrow I_3(A + 3, Z), \quad \text{etc.}$$

So long as I_N is stable, the chain of neutron captures may continue, but eventually, a radioactive isotope should be formed. Such an isotope would subsequently decay by emitting an electron (and an antineutrino), thus creating a new element:

$$I_N(A + N, Z) \rightarrow J(A + N, Z + 1) + e^- + \bar{\nu}.$$

If the new element is stable, it will resume the chain of neutron captures. Otherwise, it may undergo a series of β decays:

$$J(A + N, Z + 1) \rightarrow K(A + N, Z + 2) + e^- + \bar{\nu},$$
$$K(A + N, Z + 2) \rightarrow L(A + N, Z + 3) + e^- + \bar{\nu}, \quad \text{etc.,}$$

4.8 Creation of heavy elements: the s- and r-processes

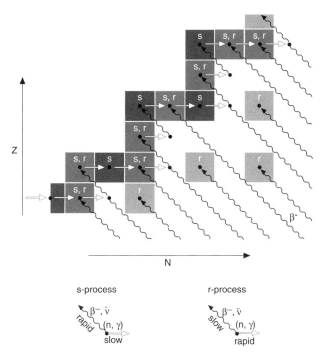

Figure 4.7 Schematic representation of the s-process and the r-process, showing reaction chains that involve neutron captures and β^- decays, leading to the formation of stable isotopes. Nuclei marked s, r, or s,r are formed by one of the processes (respectively), or by both [adapted from D. Clayton (1983), *Principles of Stellar Evolution and Nucleosynthesis*, University of Chicago Press].

until a stable nucleus of mass $\mathcal{A}+N$ and atomic number $\mathcal{Z}+M$, say, is produced. Either way, increasingly heavier elements and their stable isotopes are thereby created.

In the process just described, two types of reactions – neutron captures and β decays – and two types of nuclei – stable and unstable – are involved. Stable nuclei may, of course, undergo only neutron captures; for unstable ones both tracks are open and the outcome depends on the timescales of the two processes. The timescales of β decays (or half-life times of β-unstable isotopes) are constants – independent of prevailing physical conditions. Those of neutron captures may change according to temperature and density. Hence neutron capture reactions may proceed more *slowly* or more *rapidly* than the competing β decays. The resulting chains of reactions and products will be different in the former case, called the s-process, and in the latter, called the r-process (terms coined by Margaret and Geoffrey Burbidge, William Fowler, and Fred Hoyle in their seminal paper of 1957). This is illustrated schematically in Figure 4.7, where the s-process products are labelled s, those of the r-process are labelled r, and those which may be produced by both are labelled s,r.

In the course of the main burning processes, the s- and r-processes operate as secondary reactions and a wealth of nuclear species results, although the abundances of elements heavier than iron are relatively small.

4.9 Pair production

We have already seen in this chapter many examples of transmutations of mass into "light": all the major burning stages release energy at the expense of a small fraction (less than 1%) of the mass. But the reverse transmutation – of light into mass – is also possible.

During the interaction with a nucleus, a photon may turn into an electron-positron particle pair, provided its energy $h\nu$ exceeds the rest-mass energy of the particles, $h\nu > 2m_e c^2$. The presence of the nucleus is required for the simultaneous conservation of momentum and energy. The typical temperature at which the condition for pair production is satisfied may be estimated by $kT \approx h\nu \approx 2m_e c^2$, yielding $T \approx 1.2 \times 10^{10}$ K. However, even at temperatures $T \geq 10^9$ K, a large number of photons – at the tail of the Planck distribution function (3.39) – are already sufficiently energetic to produce electron-positron pairs. At the same time, the inverse reaction – annihilation of electron-positron pairs into photon pairs – tends to destroy the newly created positrons. As a result, the number of positrons reaches equilibrium. Pair production, as photodisintegration, bears similarity to the ionization process: an increase in temperature leads to an increase in the number of particles, at the expense of the photon energy; an increase in density has the opposite effect. Thus, at a few times 10^9 K (depending on the electron density) the number of positrons becomes a significant fraction of the number of electrons. We note that having a lot of pairs at a temperature of a few 10^9 K (much less than 12×10^9 K) is similar to having a considerable fraction of, say, ionized hydrogen at a few 10^4 K (much less than $\sim 15 \times 10^4$ K, corresponding to $\chi = 13.6$ eV).

4.10 Iron photodisintegration

If sufficiently high temperatures are achieved, even the stable iron nuclei do not survive photodisintegration. They break into α particles and neutrons,

$$^{56}\text{Fe} \rightarrow 13\,^4\text{He} + 4n,$$

thus reversing almost entirely the nucleosynthesis process. Each reaction of this kind absorbs about 100 MeV of energy.

4.10 Iron photodisintegration

At temperatures above $\sim 7 \times 10^9$ K, helium becomes more abundant than iron. At still higher temperatures, helium itself is disintegrated by the energetic photons into protons and neutrons. In conclusion, bound nuclei require temperatures above a few 10^6 K in order to be created and below a few 10^9 K so as not to be destroyed. This, as we shall see, is precisely the range of temperatures characteristic of stellar interiors.

5

Equilibrium stellar configurations – simple models

5.1 The stellar structure equations

The main conclusion of Chapter 2 was that the evolution of a star may be perceived as a quasi-static process, in which the composition changes slowly, allowing the star to maintain hydrostatic equilibrium and, generally, thermal equilibrium as well. The chain of processes through which the composition gradually changes has been described in the previous chapter. Our present task is to describe the equilibrium structure of a star of a given composition (this chapter) and to find whether the equilibrium is stable (next chapter). The (static) structure of a star is obtained from the solution of the set of differential equations known as the *stellar structure equations*, formulated in terms of either of the previously encountered space variables, r or m:

$$\frac{dP}{dr} = -\rho \frac{Gm}{r^2} \qquad \frac{dP}{dm} = -\frac{Gm}{4\pi r^4} \qquad (5.1)$$

$$\frac{dm}{dr} = 4\pi r^2 \rho \qquad \frac{dr}{dm} = \frac{1}{4\pi r^2 \rho} \qquad (5.2)$$

$$\frac{dT}{dr} = -\frac{3}{4ac} \frac{\kappa \rho}{T^3} \frac{F}{4\pi r^2} \qquad \frac{dT}{dm} = -\frac{3}{4ac} \frac{\kappa}{T^3} \frac{F}{(4\pi r^2)^2} \qquad (5.3)$$

$$\frac{dF}{dr} = 4\pi r^2 \rho q \qquad \frac{dF}{dm} = q, \qquad (5.4)$$

where the first is the hydrostatic equilibrium equation, the second is the continuity equation, the third is the radiative transfer equation (provided radiative diffusion constitutes the only means of energy transfer), and the fourth is the thermal equilibrium equation, supplemented by the relations

$$P = \frac{\mathcal{R}}{\mu_\mathrm{I}} \rho T + P_\mathrm{e} + \tfrac{1}{3} a T^4 \qquad (5.5)$$

$$\kappa = \kappa_0 \rho^a T^b \qquad (5.6)$$
$$q = q_0 \rho^m T^n. \qquad (5.7)$$

Integration of these differential equations provides the profiles of four functions throughout the star: T, ρ, m (or r), and F, from which any other function of interest may be derived. Four boundary conditions (integration constants) have to be supplied. The three straightforward ones are at $m = 0$, $r = 0$ and $F = 0$; at $m = M$ (or at $r = R$), $P = 0$. A more complicated condition relates the emitted radiation $L = F(R)$ – or, equivalently, the effective temperature – to the temperature obtained at some depth below the surface. Although this set of equations is simpler by far than the set of evolution equations derived in Chapter 2, it does not lend itself to simple, analytic solutions. The reason is threefold: first, the equations are highly nonlinear, particularly in view of the power-law relations (5.5)–(5.7); secondly, they are coupled and have to be solved simultaneously; thirdly, they constitute a two-point boundary value problem, which requires iterations for its solution. And yet, a great deal of our understanding of stellar structure dates back to the early decades of the twentieth century, when fast computers were not only unavailable, but inconceivable. Eddington's book that we have already mentioned saw light in 1926; Subrahmanyan Chandrasekhar's book *An Introduction to the Study of Stellar Structure*, a cornerstone in the study of stars, was first published in 1939.

Exercise 5.1: Derive the behaviour of $m(r)$, $P(r)$, $F(r)$, and $T(r)$ near the centre of a star by Taylor expansion for given composition and physical properties at $r = 0$: ρ_c, P_c, and T_c.

In what follows we shall see that insight into the structure of stars may be gained both by analyzing the equations, without actually solving them, and by seeking simple solutions based on additional simplifying assumptions.

5.2 What is a simple stellar model?

A fundamental principle that enables a simple solution of the structure equations is finding a property that changes moderately enough from the stellar centre to the surface to allow us to regard it as uniform (independent of r or m). At first sight, this demand appears rather strange, keeping in mind that the temperature, for example, is expected to change throughout a star by more than 3 orders of magnitude (according to simple estimates) and the pressure by more than 14! However, properties can be found that do not change significantly with radial distance. Many models, for example, assume the composition to be uniform. Is such an assumption justified? It would be for a star which is thoroughly mixed by convection (a process that we shall address shortly), or for a star composed mainly of elements heavier than hydrogen, where the gas pressure is dominated by electrons and hence depends on μ_e, which is very nearly 2 regardless of the detailed abundances. A homogeneous composition is also typical of young stars, since the initial stellar composition is uniform.

Another principle that enables an analytic investigation of the behaviour of stars is the representation of a star by its two extreme points – the centre and the surface (the surface is, of course, not a point in the strict sense of the word, but all points on the surface are identical by the spherical symmetry assumption). The hidden implication is that properties change monotonically between these two points. This is certainly correct for the pressure, from equation (5.1), and also for the temperature, by equation (5.3), since from equation (5.4), $F \geq 0$. The latter condition is not necessarily correct in the case of strong neutrino emission, which may turn the net q negative and may eventually lead to a temperature inversion. But we shall disregard such complications.

As a further simplification, we may represent a star by only one of the extreme points; the centre, for example. Assuming that both P and T decrease outward (and so must ρ; otherwise we would encounter the unstable situation in which heavy material lies on top of light material, resulting in a turnover), the centre of a star is the hottest and densest place. There, therefore, the nuclear reactions are fastest and, since nuclear processes dictate the evolutionary pace, the centre would be the most evolved part of the star. We should be able to learn a great deal about the evolution of a star by considering its central point alone. This will be the subject of Chapter 7. The surface of the star (the global stellar characteristics) is important from an entirely different point of view – it is the only "point" whose model-derived properties can be directly compared with observations. In some cases, global quantities and relations between them may be obtained, as we shall see in Chapter 7, without solving the set of structure equations.

For now, we shall consider several simple models based on the principle of a uniform property.

5.3 Polytropic models

The first pair of stellar structure equations, (5.1)–(5.2), is linked to the second pair, (5.3)–(5.4), by the dependence of pressure on temperature. If the pressure were only a function of density (and composition, of course), the first pair would be independent and could be solved separately, meaning that the hydrostatic configuration would be independent of the flow of heat through it. Analytic solutions of this form are more than a century old.

Multiplying equation (5.1) by r^2/ρ and differentiating with respect to r, we have

$$\frac{d}{dr}\left(\frac{r^2}{\rho}\frac{dP}{dr}\right) = -G\frac{dm}{dr}. \tag{5.8}$$

5.3 Polytropic models

Substituting equation (5.2) on the right-hand side, we obtain

$$\frac{1}{r^2}\frac{d}{dr}\left(\frac{r^2}{\rho}\frac{dP}{dr}\right) = -4\pi G\rho. \tag{5.9}$$

We now consider equations of state of the form

$$P = K\rho^\gamma, \tag{5.10}$$

where K and γ are constants, known as *polytropic equations of state*. It is customary to define the corresponding *polytropic index*, denoted by n, as

$$\gamma = 1 + \frac{1}{n}. \tag{5.11}$$

Thus the equation of state of a completely degenerate electron gas is polytropic, with an index of 1.5 ($\gamma = 5/3$) in the nonrelativistic case and 3 ($\gamma = 4/3$) in the extreme relativistic limit. An ideal gas, too, may be described by a polytropic equation of state under certain conditions; we shall encounter such cases later on. Substituting equations (5.10)–(5.11) into equation (5.9), we obtain a second-order differential equation:

$$\frac{(n+1)K}{4\pi Gn}\frac{1}{r^2}\frac{d}{dr}\left(\frac{r^2}{\rho^{\frac{n-1}{n}}}\frac{d\rho}{dr}\right) = -\rho. \tag{5.12}$$

The solution $\rho(r)$ for $0 \leq r \leq R$, called a *polytrope*, requires two boundary conditions. These are $\rho = 0$ at the surface ($r = R$), which follows from $P(R) = 0$, and $d\rho/dr = 0$ at the centre ($r = 0$), since hydrostatic equilibrium implies $dP/dr = 0$ there (see Section 2.3). Hence a polytrope is uniquely defined by three parameters: K, n, and R, and it enables the calculation of additional quantities as functions of radius, such as the pressure, the mass, or the gravitational acceleration.

It is convenient to define a dimensionless variable θ in the range $0 \leq \theta \leq 1$ by

$$\rho = \rho_c \theta^n, \tag{5.13}$$

to obtain equation (5.12) in a simpler form,

$$\left[\frac{(n+1)K}{4\pi G\rho_c^{\frac{n-1}{n}}}\right]\frac{1}{r^2}\frac{d}{dr}\left(r^2\frac{d\theta}{dr}\right) = -\theta^n. \tag{5.14}$$

Obviously, the coefficient in square brackets on the left-hand side of equation

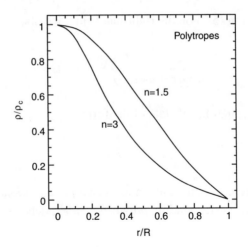

Figure 5.1 Normalized polytropes for $n = 1.5$ and $n = 3$.

(5.14) is a constant having the dimension of length squared,

$$\left[\frac{(n+1)K}{4\pi G \rho_c^{\frac{n-1}{n}}}\right] = \alpha^2, \tag{5.15}$$

which can be used in order to replace r by a dimensionless variable ξ,

$$r = \alpha\xi. \tag{5.16}$$

Substituting equation (5.16) into equation (5.14), we now obtain the well-known Lane–Emden equation of index n,

$$\frac{1}{\xi^2}\frac{d}{d\xi}\left(\xi^2\frac{d\theta}{d\xi}\right) = -\theta^n, \tag{5.17}$$

subject to the boundary conditions: $\theta = 1$ and $d\theta/d\xi = 0$ at $\xi = 0$. Equation (5.17) can be integrated starting at $\xi = 0$; for $n < 5$, the solutions $\theta(\xi)$ are found to decrease monotonically and have a zero at a finite value $\xi = \xi_1$, which corresponds to the stellar radius,

$$R = \alpha\xi_1. \tag{5.18}$$

Examples of solutions (ρ/ρ_c as a function of r/R), for $n = 1.5$ and $n = 3$, are given in Figure 5.1. As shown, the structure of a polytrope depends only on n. A polytrope of index 3 describes a star in which the mass is strongly concentrated at the centre, whereas a polytrope of index 1.5 describes a more even mass distribution.

5.3 Polytropic models

Table 5.1 Polytropic constants

n	D_n	M_n	R_n	B_n
1.0	3.290	3.14	3.14	0.233
1.5	5.991	2.71	3.65	0.206
2.0	11.40	2.41	4.35	0.185
2.5	23.41	2.19	5.36	0.170
3.0	54.18	2.02	6.90	0.157
3.5	152.9	1.89	9.54	0.145

The total mass M of a polytropic star is given by

$$M = \int_0^R 4\pi r^2 \rho \, dr = 4\pi \alpha^3 \rho_c \int_0^{\xi_1} \xi^2 \theta^n d\xi. \tag{5.19}$$

From equation (5.17) we have

$$M = -4\pi \alpha^3 \rho_c \int_0^{\xi_1} \frac{d}{d\xi}\left(\xi^2 \frac{d\theta}{d\xi}\right) d\xi = -4\pi \alpha^3 \rho_c \xi_1^2 \left(\frac{d\theta}{d\xi}\right)_{\xi_1}. \tag{5.20}$$

Exercise 5.2: Solve the Lane–Emden equation analytically for (a) $n = 0$ and (b) $n = 1$ and find ξ_1 and $M(R)$ in each case.

In later discussions we shall often resort to general relations between stellar properties resulting from a polytropic equation of state. These follow easily from equation (5.20). Eliminating α between equations (5.18) and (5.20), we obtain a linear relation between the central density and the average density $\bar{\rho}$,

$$\rho_c = D_n \bar{\rho} = D_n \frac{M}{\frac{4\pi}{3}R^3}, \tag{5.21}$$

which is generally valid. Only the constant D_n derives from the solution of equation (5.17) and depends on the value of n:

$$D_n = -\left[\frac{3}{\xi_1}\left(\frac{d\theta}{d\xi}\right)_{\xi_1}\right]^{-1}. \tag{5.22}$$

Values of D_n for various n can be found in Table 5.1.

Using equation (5.20) again, but now eliminating ρ_c with the aid of equation (5.15) and substituting α from equation (5.18), we obtain a relation between the stellar mass and radius, which may be expressed in terms of two constants, M_n

and R_n, in the form

$$\left(\frac{GM}{M_n}\right)^{n-1}\left(\frac{R}{R_n}\right)^{3-n} = \frac{[(n+1)K]^n}{4\pi G}. \tag{5.23}$$

The values of the constants $M_n = -\xi_1^2(d\theta/d\xi)_{\xi_1}$ and $R_n = \xi_1$ vary with the polytropic index n in the range from 1 to 10, as listed in Table 5.1. We note that $n = 3$ is a special case: the mass becomes independent of radius and is uniquely determined by K,

$$M = 4\pi M_3 \left(\frac{K}{\pi G}\right)^{3/2}. \tag{5.24}$$

Thus for a given K, there is only one possible value for the mass of a star that will satisfy hydrostatic equilibrium. Another special case is $n = 1$, for which the radius is independent of mass and is uniquely determined by K:

$$R = R_1 \left(\frac{K}{2\pi G}\right)^{1/2}. \tag{5.25}$$

Between these limiting values of n, $1 < n < 3$, we have from equation (5.23)

$$R^{3-n} \propto \frac{1}{M^{n-1}}, \tag{5.26}$$

meaning that the radius decreases with increasing mass: the more massive the star, the smaller (and hence denser) it becomes.

A final important relation is obtained between the central pressure and the central density by substituting K from the mass-radius relation (5.23) in equation (5.10), $P_c = K\rho_c^{1+\frac{1}{n}}$, whence

$$P_c = \frac{(4\pi G)^{\frac{1}{n}}}{n+1} \left(\frac{GM}{M_n}\right)^{\frac{n-1}{n}} \left(\frac{R}{R_n}\right)^{\frac{3-n}{n}} \rho_c^{\frac{n+1}{n}}. \tag{5.27}$$

Eliminating R between equations (5.27) and (5.21), and assembling all n-dependent coefficients into one constant B_n, reduces equation (5.27) to

$$P_c = (4\pi)^{1/3} B_n G M^{2/3} \rho_c^{4/3}. \tag{5.28}$$

The remarkable property of this relation is that it depends on the polytropic equation of state *only* through the value of B_n, which, as we see from Table 5.1, varies very slowly with n. It therefore constitutes an almost universal relation, and as such it will be used in Chapter 7. Note that expression (5.28) for P_c is consistent with the upper limit derived in Exercise 2.2 (Section 2.3).

> **Exercise 5.3:** For a given mass M and central pressure P_c, which polytrope yields a bigger star: that of index 1.5 or that of index 3?
>
> **Exercise 5.4:** *Capella* is a binary star discovered in 1899, with a known orbital period, which enables the determination of the mass and radius of the brighter component: $M = 8.3 \times 10^{30}$ kg and $R = 9.55 \times 10^9$ m. Assuming that the star can be described by a polytrope of index 3, find the central pressure and the central density. Check whether the central pressure satisfies inequality (2.18).

5.4 The Chandrasekhar mass

Stars that are so dense as to be dominated by the degeneracy pressure of the electrons (discussed in Chapter 3) would be accurately described by a polytrope of index $n = 1.5$, with $K = K_1$ of equation (3.35). We know from observations that such compact stars exist – they are the *white dwarfs* mentioned in Chapter 1, which have masses comparable to the Sun's and radii not much larger than the Earth's. Their average density is thus higher than 10^8 kg m^{-3} (10^5 g cm^{-3}), about five orders of magnitude higher than the average density of the Sun. We might learn some more about these stars by investigating the properties of this particular polytrope. From equation (5.23), the relation between mass and radius becomes

$$R \propto M^{-1/3}. \tag{5.29}$$

The density, therefore, increases as the square of the mass,

$$\bar{\rho} \propto MR^{-3} \propto M^2. \tag{5.30}$$

Imagine now a series of such degenerate gaseous spheres with higher and higher masses. The radii will decrease along the series and the density will increase in proportion to M^2. Eventually, the density will become so high that the degenerate electron gas will turn to be relativistic, departing from the simple $n = 1.5$ polytrope. As the density increases (the radius tending to zero), the correct equation of state will approach the form (3.38), still a polytrope, but of index $n = 3$, with $K = K_2$. We have seen, however, that in such a case there is only one possible solution for M, uniquely determined by K. Hence our series of degenerate gaseous spheres in hydrostatic equilibrium ends at this limiting mass. The existence of

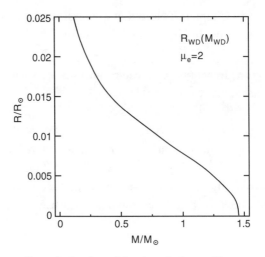

Figure 5.2 The mass-radius relation for white dwarfs ($\mu_e = 2$).

an upper limit to the mass of degenerate stars was first found by Chandrasekhar in 1931, and hence the upper limit bears his name, M_{Ch}. About half a century later, this work earned Chandrasekhar the 1983 Nobel Prize for Physics, which he shared with Fowler (for their contributions to the understanding of stellar evolution).

Substituting K_2 in equation (5.24), we have

$$M_{Ch} = \frac{M_3 \sqrt{1.5}}{4\pi} \left(\frac{hc}{Gm_H^{4/3}} \right)^{3/2} \mu_e^{-2}. \qquad (5.31)$$

Inserting the values of constants, we obtain

$$M_{Ch} = 5.83 \mu_e^{-2} M_\odot, \qquad (5.32)$$

which yields for $\mu_e = 2$ a limiting mass of $1.46 M_\odot$. The mass-radius relation for white dwarfs is shown in Figure 5.2 for $\mu_e = 2$ (He, C, O, ...). For $\mu_e = 2.15$ (Fe), the limiting mass is $1.26\ M_\odot$. In conclusion, hydrogen-poor compact stars, where the pressure is supplied predominantly by the degenerate electron gas, can have masses only up to the critical mass of $1.46 M_\odot$. Indeed, no white dwarf is known with a mass exceeding this value.

Exercise 5.5: Calculate the critical mass using relation (5.28) between central pressure and central density; show that the numerical coefficient in equation (5.31) is equivalent to $B_3^{-3/2} \sqrt{1.5/32\pi}$.

5.5 The Eddington luminosity

So far we have dealt with the first two of the structure equations. We shall now add the third, thus taking into account the temperature and the radiation pressure. Substituting the radiation pressure $P_{\rm rad} = \frac{1}{3}aT^4$ in equation (5.3) and dividing equation (5.3) by equation (5.1), we obtain

$$\frac{dP_{\rm rad}}{dP} = \frac{\kappa F}{4\pi cGm}. \tag{5.33}$$

The result of this manipulation will be the derivation of an upper limit for the stellar luminosity. Since $P = P_{\rm gas} + P_{\rm rad}$, and both $P_{\rm gas}$ and $P_{\rm rad}$ decrease outward (provided $q \geq 0$), it follows that $\text{sgn}[dP_{\rm rad}] = \text{sgn}[dP_{\rm gas}]$ and we obviously have $dP_{\rm rad}/dP < 1$, implying

$$\kappa F < 4\pi cGm. \tag{5.34}$$

This inequality may be violated either in the case of a very large heat flux, which may result from intense nuclear burning, or in the case of a very high opacity, as encountered at the ionization temperatures of hydrogen or helium. In such cases equations (5.1) and (5.3) cannot simultaneously hold, and if we require hydrostatic equilibrium, then heat transport must be described by a different equation; that is, it must occur by a means other than radiative diffusion, which has become inefficient. We know from everyday experience that near a strong heat source, such as a stove, convective motions develop in the surrounding air, which carry the heat efficiently and distribute it throughout the room. If the stove is not very hot, it spreads heat by thermal radiation alone. The same phenomenon occurs in stars – on appropriately larger scales. Stars transfer energy by radiation alone under moderate conditions, in which case they are said to be in *radiative equilibrium* and inequality (5.34) is satisfied, or by convection, under more severe conditions – when the rate of heat generation becomes too rapid for radiation to carry, or when ionization interferes too much with the radiative transfer. It may also happen that some regions of a star are in radiative equilibrium and others are not; the former are called *radiative* regions or zones, and the latter are called *convective* ones.

Exercise 5.6: Find the expression for the gas pressure gradient, assuming radiative equilibrium, and its relation to equation (5.34).

Near the centre of a star, equation (5.4) and $F(0) = 0$ yield $F/m \to q_c$ as $m \to 0$, where $q_c = q(m = 0)$; hence inequality (5.34) imposes a universal upper limit on the central energy generation rate that can be accommodated by radiative energy transfer:

$$q_c < \frac{4\pi cG}{\kappa}. \tag{5.35}$$

The surface layer of a star is always radiative; applying inequality (5.34) for $m = M$, we have

$$L < \frac{4\pi cGM}{\kappa}. \tag{5.36}$$

Violation of this condition then implies violation of hydrostatic equilibrium: mass motions arise leading to a stellar wind. As pointed out by Eddington, the right-hand side of inequality (5.36) represents a critical luminosity L_{crit} that cannot be surpassed; it is, therefore, also known as the Eddington luminosity L_{Edd}:

$$L_{\text{Edd}} = \frac{4\pi cGM}{\kappa} = 3.2 \times 10^4 \left(\frac{M}{M_\odot}\right)\left(\frac{\kappa_{\text{es}}}{\kappa}\right) L_\odot, \tag{5.37}$$

where the opacity is expressed relative to the electron-scattering opacity κ_{es}, which is a constant [see equation (3.64)]. To summarize, radiative equilibrium requires

$$L < L_{\text{Edd}}.$$

To show the possible implications of this result, we may indulge in some speculation. If we assume $\kappa \approx \kappa_{\text{es}}$ to be a reasonable approximation, L_{Edd} becomes uniquely determined by M. We have seen in Chapter 1 that for a certain type of stars, those of the *main sequence*, a correlation exists between the luminosity and the mass. If the outer layers of such stars are in hydrostatic and radiative equilibrium, restriction (5.36) combined with the mass-luminosity relation imposes an upper limit on the mass of main-sequence stars. We should then expect the main sequence to have an upper end.

5.6 The standard model

After this brief digression, we proceed to derive the so-called *standard* model, which is due to Eddington and is therefore also known as *Eddington's model*.

We define a function η by

$$\frac{F}{m} = \eta \frac{L}{M} \tag{5.38}$$

5.6 The standard model

and insert it into equation (5.33), which becomes

$$\frac{dP_{\text{rad}}}{dP} = \frac{L}{4\pi cGM}\kappa\eta. \tag{5.39}$$

At the surface, $\eta = 1$, and for stars that burn nuclear fuel mostly in a (small) central core, thus maintaining an almost constant flux outside the core, η increases inward, as m decreases. The opacity, on the other hand, usually increases from the centre outward. If, from the centre outward, the increase in κ is approximately compensated by the decrease in η, we may take their product to be constant. This is the uniform property of the Eddington model (a controversial assumption, which has been subject to severe criticism over the years). With

$$\kappa\eta = \text{constant} \equiv \kappa_s, \tag{5.40}$$

where κ_s is the surface opacity, we have by integrating equation (5.39)

$$P_{\text{rad}} = \frac{\kappa_s L}{4\pi cGM} P, \tag{5.41}$$

since the total pressure and the radiation pressure tend to zero at the surface. Thus the constancy of $\kappa\eta$ implies a constant ratio of radiation pressure to total pressure throughout the star; in other words, a constant β [see equation (3.12)]. We also obtain

$$L = \frac{4\pi cGM}{\kappa_s}(1-\beta) = L_{\text{Edd}}(1-\beta), \tag{5.42}$$

meaning that the luminosity approaches the limiting value as the radiation pressure becomes dominant ($\beta \to 0$). Assuming the gas pressure to be given by the ideal gas law, equation (3.28), we have

$$\frac{aT^4}{3(1-\beta)} = \frac{P_{\text{rad}}}{1-\beta} = P = \frac{P_{\text{gas}}}{\beta} = \frac{\mathcal{R}}{\beta\mu}\rho T. \tag{5.43}$$

Combining the extreme left and extreme right expressions, we get

$$T = \left[\frac{3\mathcal{R}(1-\beta)}{a\mu\beta}\right]^{1/3}\rho^{1/3}, \tag{5.44}$$

and the equation of state may be written as

$$P = K\rho^{4/3} \qquad K = \left[\frac{3\mathcal{R}^4(1-\beta)}{a\mu^4\beta^4}\right]^{1/3}. \tag{5.45}$$

Since K is a constant, we have obtained a polytropic equation of state of index 3,

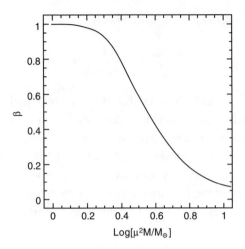

Figure 5.3 Solution of the Eddington quartic equation.

which implies a unique relation between K and M, equation (5.24) derived in the previous section. Rearranging terms and inserting the values of constants, we obtain

$$1 - \beta = 0.003 \left(\frac{M}{M_\odot}\right)^2 \mu^4 \beta^4, \tag{5.46}$$

a fourth-order equation for $\beta(\mu^2 M)$, known as the *Eddington quartic equation*, whose solution is given in Figure 5.3. The quartic equation is valid for a hypothetically wide range of masses. We note, however, that only for a rather restricted range does β differ significantly from unity (pure gas pressure) or from zero (pure radiation pressure), and this range more or less coincides with the range of stellar masses, as derived from observations. In Eddington's own words,

> "We can imagine a physicist on a cloud-bound planet who has never heard tell of the stars calculating the ratio of radiation pressure to gas pressure for a series of globes of gas of various sizes, starting, say, with a globe of mass 10 gm., then 100 gm., 1000 gm., and so on, so that his nth globe contains 10^n gm. ... Regarded as a tussle between matter and aether (gas pressure and radiation pressure) the contest is overwhelmingly one-sided except between Nos. 33-35, where we may expect something interesting to happen.
> What 'happens' is the stars.
> We draw aside the veil of cloud beneath which our physicist has been working and let him look up at the sky. There he will find a thousand million globes of gas nearly all of mass between his 33rd and 35th globes – that is to say, between $\frac{1}{2}$ and 50 times the sun's mass."
>
> Sir Arthur S. Eddington: The Internal Constitution of the Stars, 1926

5.6 The standard model

Exercise 5.7: The quartic equation may be written in terms of a mass $M_\star$ that is a combination of natural constants. (a) Find the expression for this mass and calculate it. (b) Express the Chandrasekhar mass in terms of $M_\star$.

What can one learn about the evolution of stars based on this simple model?

1. For stars of given composition (fixed μ), β decreases as M increases, meaning that radiation pressure becomes particularly important in massive stars.
2. Inserting equation (5.42) into equation (5.46), we obtain

$$\frac{L}{L_\odot} = \frac{4\pi c G M_\odot}{\kappa_s L_\odot} 0.003 \mu^4 \beta(M,\mu)^4 \left(\frac{M}{M_\odot}\right)^3. \qquad (5.47)$$

This is close to a power-law relation between the luminosity and the mass, similar to that obtained from observations of main-sequence stars. In fact, the mass-luminosity relation derived by Eddington was at the time a theoretical prediction, to be confirmed only later by observations. Differences in composition, and hence in the value of μ, may explain the observed scatter of points in the (M, L) relation.

3. For a given M and changing μ – as along the evolutionary course of a star – β decreases with increasing μ. Since nuclear reactions cause a gradual increase in μ, we should expect radiation pressure to play an increasingly greater role as a star gets older. With it, by equation (5.42), the luminosity should approach its limiting value. Could this mean that a star should lose (eject) some of its mass in its late stages of evolution, compensating for the rise in μ, so as to prevent β from dropping too low? We may consider this a first hint to the existence of stellar winds, which should intensify as the luminosity approaches L_{Edd}.

We have formulated these conclusions very cautiously, for they derive from such a simple model. So formulated, they are acceptable and they provide important and easy to understand clues to the complex structure and evolution of stars.

Historical Note: The theoretical mass-luminosity relation (5.47) has two parameters: the mean molecular weight and the opacity. Assuming one of them, one may derive the other by comparing the relation with its observational counterpart. Eddington started by assuming stars to be made of iron (or terrestrial material), which implied a value of μ slightly in excess of 2, considering the highly ionized state of stellar interiors. This led to the estimate of an "astronomical opacity coefficient," which exceeded by about a factor of 10 the "theoretical opacity coefficient" that had been calculated following the Kramers theory. Although he was aware

> that including a considerable proportion of hydrogen in the chemical composition of stars would resolve the discrepancy, this solution seemed improbable at first, both to him and to others, and the alternative of seeking a correction to the opacity coefficient was pursued for a time. However, around 1930 it became established that in the atmospheres of the Sun, and the stars in general, hydrogen amounts to about half the mass. The possibility of hydrogen floating to the surface of the star was discarded; Eddington had already shown that diffusion in stars should proceed negligibly slowly. Thus, in 1932, the prevalence of hydrogen in stellar interiors was finally recognized by Eddington, and independently advocated by Bengt Strömgren, on the basis of the mass-luminosity relation.

5.7 The point-source model

Models presented so far have considered the first three structure equations. This chapter would not be complete without mentioning another group of relatively simple models, which take account of the fourth equation, equation (5.4), and assume a power law for the opacity [in the form (5.6)]. If the nuclear energy source of a star is confined to a very small central region, it may be considered a *point source*, so that $q = 0$ for $r > 0$. In this case, the equation of thermal equilibrium implies a constant energy flow (energy per unit time) throughout the star. Thus

$$F = \text{constant} = F(R) \equiv L \tag{5.48}$$

constitutes the basic assumption of the point-source models. Such models were first investigated by Thomas Cowling, in 1930, and hence they are also known as *Cowling models*. It is reasonable to assume a homogeneous composition for point-source models. Expressing the opacity in terms of ρ and T as in equation (3.63), $\kappa = \kappa_0 \rho^a T^b$, the set of equations to be solved reduces to

$$\frac{dP}{dr} = -\frac{Gm\rho}{r^2} \tag{5.49a}$$

$$\frac{dP_{\text{rad}}}{dr} = -\frac{\kappa_0 L}{c} \frac{\rho^{a+1} T^b}{4\pi r^2} \tag{5.49b}$$

$$\frac{dm}{dr} = 4\pi r^2 \rho, \tag{5.49c}$$

together with an equation of state for the gas, say $P_{\text{gas}} = (\mathcal{R}/\mu)\rho T$. This is by no means a very simple or transparent model, but equations (5.49) can be integrated numerically for a given opacity law. A somewhat simpler and more elegant version of the point-source model may be obtained if one further assumes the opacity to be constant ($a = b = 0$). Equations (5.49a) and (5.49b) may then be written as

$$\frac{d(P_{gas} + P_{rad})}{dr} = -\frac{Gm\rho}{r^2} \quad (5.50a)$$

$$\frac{dP_{rad}}{dr} = -\frac{\kappa L \rho}{4\pi c r^2}, \quad (5.50b)$$

and dividing them [as we did to obtain equation (5.33)], we get

$$\frac{dP_{gas}}{dP_{rad}} = \frac{4\pi c G}{\kappa L} m - 1. \quad (5.51)$$

We now differentiate equation (5.51) to obtain

$$\frac{d^2 P_{gas}}{dP_{rad}^2} \frac{dP_{rad}}{dr} = \frac{4\pi c G}{\kappa L} \frac{dm}{dr}. \quad (5.52)$$

Inserting equations (5.50b) on the left-hand side and (5.49c) on the right-hand side, and rearranging terms, we finally have

$$\frac{d^2 P_{gas}}{dP_{rad}^2} = -\left(\frac{64\pi^3 c^2 G}{\kappa^2 L^2}\right) r^4. \quad (5.53)$$

We may express ρ in terms of P_{gas} and P_{rad} as follows:

$$\rho = \frac{\mu}{\mathcal{R}} \left(\frac{a}{3}\right)^{1/4} P_{gas} P_{rad}^{-1/4}, \quad (5.54)$$

and cast equation (5.50b) in the form

$$\frac{d}{dP_{rad}}\left(\frac{1}{r}\right) = \frac{4\pi c \mathcal{R}}{\mu \kappa L}\left(\frac{3}{a}\right)^{1/4} P_{gas}^{-1} P_{rad}^{1/4}. \quad (5.55)$$

Thus the original set of four equations has been reduced to a pair of differential equations, (5.53) and (5.55), in three variables: P_{gas}, r, and P_{rad} as the independent one. With the introduction of appropriate dimensionless variables – x for P_{rad}, y for P_{gas}, and z for r^{-1} – the equations to be solved become

$$\frac{d^2 y}{dx^2} = -z^{-4} \qquad \frac{dz}{dx} = y^{-1} x^{1/4}. \quad (5.56)$$

The solutions $y(x)$ and $z(x)$ may be inverted to obtain $P_{gas}(r)$ and $P_{rad}(r)$, and with them the temperature and density variation throughout the star. Even the solution of this pair of equations is far from being straightforward (a detailed analysis may be found in Chandrasekhar's book, *An Introduction to the Study of Stellar*

Structure, 1939). It is interesting to note that the point-source model yields a mass-luminosity relation whose slope, on logarithmic scales, is much less steep for large masses than for small ones, which is in qualitative agreement with observations (Figure 1.6). A tendency toward a steeper slope for low mass stars is exhibited by the relation resulting from the standard model [equation (5.47)] as well.

Note: We stress the mass-luminosity relation in particular, because of the primary role it played at the early stage of the stellar evolution theory, when a great deal of confusion regarding the nuclear reactions responsible for energy generation still prevailed:

"Our discussion has been based on the relation generally called the mass-luminosity-relation. ... The relation however contains several unknowns, and without certain assumptions with regard to some of them no definite results can be reached. So far it is the only relation between the unknowns in question, which it has been possible to establish. When our knowledge of the energy-generation in the stars advances so that one more relation can be established, we shall probably be able to give definite answers to the questions raised by the discussion."
Bengt Strömgren: On the Interpretation of the Hertzsprung-Russell-Diagram, in
Zeitschrift für Astrophysik, 1933

And, a few years later,

"The researches of the last two decades into the constitution of the stars have resulted in considerable advance in the understanding of the physical processes in stellar interiors. The chief success of the investigations is the establishing of a mass-luminosity relation. This relation has been obtained without reference to the actual nuclear reactions that are the source of stellar energy, merely from consideration of the mechanical and thermodynamical equilibrium of the star. It follows therefore that exact knowledge of the rate of generation of subatomic energy cannot overthrow the mass-luminosity relation, but may serve only to place some restriction on the range of magnitude [luminosity] corresponding to a given mass."
Fred Hoyle and Raymond A. Lyttleton: The Evolution of the Stars, in
Proceedings of the Cambridge Philosophical Society, 1939

When nuclear energy generation in stars finally became understood, Cowling wrote in retrospection:

"With the advent of this new physical information, complete data for constructing stellar models were for the first time available; it was like supplying the fourth leg of a chair which so far had only one back leg."
Thomas G. Cowling: The Development of the Theory of Stellar Structure,
Quarterly Journal of the Royal Astronomical Society, 1966

Models such as those briefly mentioned here were developed more than 60 years ago, long before computers became available. Since the advent of computers they became rather obsolete, for the relatively simple computation involved ceased to be a real advantage. They have been brought here mainly for one purpose: to demonstrate how complicated is the solution of the apparently simple set of structure equations, even under the most extreme simplifying assumptions that are still (barely) consistent with physical reality.

6

The stability of stars

In the previous chapter we have dealt with models of the stellar structure under conditions of thermal and hydrostatic equilibrium. But in order to accomplish our first task toward understanding the process of stellar evolution – the investigation of equilibrium configurations – we must test the equilibrium configurations for stability. The difference between stable and unstable equilibrium is illustrated in Figure 6.1 by two balls: one on top of a dome and the other at the bottom of a bowl. Obviously, the former is in an unstable equilibrium state, while the latter is in a stable one. The way to prove (or test) this statement is also obvious and it is generally applicable; it involves a small perturbation of the equilibrium state. Imagine the ball to be slightly perturbed from its position, resulting in a slight imbalance of the forces acting on it. In the first case, this would cause the ball to slide down, running away from its original position. In the second case, on the other hand, the perturbation will lead to small oscillations around the equilibrium position, which friction will eventually dampen, the ball thus returning to its original point. The small imbalance *led* to the restoration of equilibrium by opposing the tendency of the perturbation. Thus a stable equilibrium may be maintained indefinitely, while an unstable one must end in a runaway, for random small perturbations are always to be expected in realistic physical systems.

As stars preserve their properties for very long periods of time, we may guess that their state of equilibrium is stable. But is it always? What is the mechanism that renders it stable? Is this mechanism always operating? If not, what are the conditions required for it to operate? We shall presently address these questions for each of the two types of equilibrium of stellar configurations: thermal and hydrostatic.

6.1 Secular thermal stability

The total energy of a star in hydrostatic equilibrium is given by the sum of the internal energy U and the gravitational potential energy Ω, as we have seen in

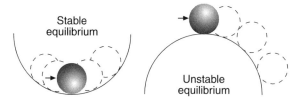

Figure 6.1 Illustration of stable (left) and unstable (right) equilibrium states.

Chapter 2. These are related by the virial theorem, equation (2.23):

$$3 \int_0^M \frac{P}{\rho} dm = -\Omega.$$

In the case of an ideal gas with negligible radiation pressure, we have by equation (3.44)

$$U = -\tfrac{1}{2}\Omega \qquad (6.1)$$

and consequently,

$$E = \tfrac{1}{2}\Omega = -U. \qquad (6.2)$$

For an ideal gas and a nonnegligible radiation pressure, we have from equations (3.28), (3.40), (3.44), and (3.47)

$$\frac{P}{\rho} = \frac{P_{\text{gas}}}{\rho} + \frac{P_{\text{rad}}}{\rho} = \frac{\mathcal{R}}{\mu}T + \frac{aT^4}{3\rho} = \tfrac{2}{3}u_{\text{gas}} + \tfrac{1}{3}u_{\text{rad}}.$$

Applying the virial theorem, we obtain

$$U_{\text{gas}} = -\tfrac{1}{2}(\Omega + U_{\text{rad}}), \qquad (6.3)$$

where U_{gas} is the total internal energy of the gas and U_{rad} is the total radiation energy, whence

$$E = \tfrac{1}{2}(\Omega + U_{\text{rad}}) = -U_{\text{gas}}. \qquad (6.4)$$

The effect of radiation is to reduce the gravitational attraction; $\Omega + U_{\text{rad}}$ may thus be regarded as an effective gravitational potential energy. In both cases the star heats up upon contraction (| Ω | increases and with it U_{gas}, and hence the average temperature) and cools upon expansion.

We have also seen that the rate of change of the energy is given by the difference between the rate of nuclear energy production and the rate of emission of radiation:

$$\dot{E} = L_{\text{nuc}} - L. \tag{6.5}$$

A state of thermal equilibrium is obtained when these terms are in balance, $L_{\text{nuc}} = L$, and hence the energy is constant ($\dot{E} = 0$). Suppose now that a small perturbation causes a slight imbalance, so that L_{nuc} exceeds L. By equation (6.5), the total energy will increase ($\dot{E} > 0$), and since E is negative, it means that its absolute value will become smaller. Therefore, by equation (6.2) or (6.4), the average temperature will decrease. At the same time the star will expand, the average density thus decreasing. As a result, the (average) rate of nuclear reactions, which is proportional to positive powers of ρ and T, will slow down and L_{nuc} will drop. The perturbation will be reversed and thermal equilibrium will eventually be restored. This thermostat, provided by the virial theorem, is the stabilizing mechanism by which stars are capable of maintaining thermal equilibrium for such long times. Stars are said to be in a state of *secular stability*.

6.2 Cases of thermal instability

The crucial link in the chain of arguments leading to the conclusion of secular stability was the dependence of the internal energy of the star on temperature; more precisely, the *negative* heat capacity of stars. Only if a change in internal energy involves a change in temperature that, in turn, affects the energy supply, is the thermal stability secured.

Thermal instability of degenerate gases

We have seen that when the pressure is due mainly to the degenerate electron gas, it is insensitive to temperature. The same applies to the internal energy of the gas (as shown in Section 3.5), and hence although equation (6.2) still holds (for a nonrelativistic gas), a decrease in internal energy – resulting from a perturbation $L_{\text{nuc}} > L$ – *will* lead to expansion, but it *will not* entail a drop in temperature. Since nuclear energy production is far more sensitive to temperature than to density, the nuclear energy output will not diminish. Instead of a restoration of thermal equilibrium, a runaway from equilibrium will ensue: the temperature will continue to rise due to the enhanced nuclear energy release, this will cause the nuclear energy generation to escalate, and so forth. Such an instability is called a *thermonuclear runaway*. It is encountered whenever nuclear reactions ignite in a degenerate gas, and it may result in an explosion. A catastrophic outcome may, however, be avoided: the gas may, eventually, become sufficiently hot and diluted to behave as an ideal gas, for which the stabilizing mechanism operates. We say in this case

6.2 Cases of thermal instability

that the degeneracy has been *lifted*. An entire class of stellar outbursts – known as *novae* – constitute notorious examples of such thermonuclear runaways, which develop into explosions on the surfaces of white dwarfs and are subsequently quenched.

The secular instability caused by temperature-sensitive nuclear reactions in degenerate matter was first studied by Tsung-Dao Lee (in 1950) and Leon Mestel (in 1952). In 1958 Evry Schatzman proposed that unstable burning on the surfaces of white dwarfs may lead to recurrent ejection of gaseous shells. This avant-garde suggestion, then barely supported by observations, became in time the very model of nova outbursts.

> **Exercise 6.1:** Explosive hydrogen burning at the bottom of a thin hydrogen-rich layer on the surface of a white dwarf will eventually lead to the expulsion of this layer. (a) For a white dwarf of mass $M = M_\odot$, which has a radius $R \approx 0.01 R_\odot$, calculate the fraction f of the layer's mass that has to be transformed into helium in order to supply the energy necessary for expulsion, assuming the layer to be of solar composition. (b) Derive the dependence of f on M for $M < M_{\text{Ch}}$.

To better understand the application of the stability criterion, consider a star which burns nuclear fuel at its centre. In hydrostatic equilibrium, the central pressure and density are related by

$$\frac{dP_c}{P_c} = \frac{4}{3}\frac{d\rho_c}{\rho_c} \tag{6.6}$$

[cf. equation (5.28)]. The pressure, density, and temperature are linked by the equation of state, which may be written in general form as

$$\frac{dP_c}{P_c} = a\frac{d\rho_c}{\rho_c} + b\frac{dT_c}{T_c}, \tag{6.7}$$

where a and b are positive coefficients. Combining equations (6.6) and (6.7), we obtain

$$\left(\frac{4}{3} - a\right)\frac{d\rho_c}{\rho_c} = b\frac{dT_c}{T_c}. \tag{6.8}$$

So long as $a < 4/3$, $\text{sgn}[d\rho_c/\rho_c] = \text{sgn}[dT_c/T_c]$, and hence contraction (caused by energy loss) is accompanied by heating, while expansion (caused by energy gain) is accompanied by cooling, as required for stability. This is the case for ideal gases, where $a = b = 1$. For degenerate material, on the other hand, $a \gtrsim 4/3$ and $0 < b \ll 1$. Thus $d\rho_c/\rho_c$ and dT_c/T_c have opposite signs. This means that expansion, which would result from an increase in internal energy, would be accompanied by a (small) *rise* in temperature, that in turn would lead to a further enhancement of $L_{\rm nuc}$. Such a situation is obviously unstable; but since the temperature rises as the gas expands, the gas may gradually become ideal (in terms of the coefficients a and b, the former decreasing and the latter increasing), in which case stability would be restored. We note, in passing, that, generalizing equation (6.8), a degenerate star that loses energy is expected to contract and *cool*, unlike an ideal gas one.

The thin shell instability

Consider a thin shell of mass Δm, temperature T, and density ρ within a star of radius R, between a fixed inner boundary r_0 and an outer boundary r, so that its thickness is $\ell = r - r_0 \ll R$. Assume nuclear reactions to take place in this shell. If the shell is in thermal equilibrium, the rate of nuclear energy generation is equal to the net rate of heat flowing out of the shell [cf. equation (5.4)]. If the rate of energy generation exceeds the rate of heat flow, the shell will expand, thus pushing outward the layers above it. Lifting of these layers will result in a diminished pressure. In hydrostatic equilibrium, the pressure within the shell, determined by the weight of the layers above it, varies as r^{-4} [cf. equation (5.1)]; hence

$$\frac{dP}{P} = -4\frac{dr}{r}. \tag{6.9}$$

The shell's mass is given by $\Delta m = 4\pi r_0^2 \ell \rho$ and therefore the density varies with the shell's thickness as

$$\frac{d\rho}{\rho} = -\frac{d\ell}{\ell} = -\frac{dr}{\ell} = -\frac{dr}{r}\frac{r}{\ell}. \tag{6.10}$$

Substituting dr/r from equation (6.9), we obtain a relation between the changes in density and pressure of the form

$$\frac{dP}{P} = 4\frac{\ell}{r}\frac{d\rho}{\rho}. \tag{6.11}$$

To obtain the resulting change in temperature, we use the equation of state in the general form (6.7), leading to

$$\left(4\frac{\ell}{r} - a\right)\frac{d\rho}{\rho} = b\frac{dT}{T}. \tag{6.12}$$

For thermal stability we require the expansion of the shell to result in a drop in temperature, or, since $b > 0$,

$$4\frac{\ell}{r} > a. \tag{6.13}$$

Obviously, for a sufficiently thin shell ($\ell/r \to 0$) the stability condition would eventually be violated. If the shell is too thin, its temperature increases upon expansion (even if the gas within it is an ideal gas), and this may lead to a runaway. Thus with respect to nuclear energy generation, a thin shell behaves much in the same way as a degenerate gas. The thermal instability of thin shells was first pointed out by Martin Schwarzschild and Richard Härm in 1965.

Before we leave the subject of thermal stability (or lack thereof), a word of caution would be in order. In all our foregoing discussions we have neglected the possibility that a change in temperature might affect not only the energy generation rate, but also the heat flux. Thus an increase in temperature, resulting from $L_{\rm nuc} < L$, may not only lead to a higher $L_{\rm nuc}$, but also to a higher L, and the outcome $L_{\rm nuc} > L$ – interpreted as thermal stability – might not be guaranteed. Similarly, if a surplus of heat in a thin shell causes the temperature to rise, in spite of the shell's expansion, this might enhance the rate of heat flow out of the shell so as to prevent a runaway, even if the rate of nuclear energy generation increases. As it happens, the heat flux is far less sensitive to temperature than is the rate of nuclear energy generation. Hence, generally, changes in L or dF/dm may safely be neglected compared with changes in $L_{\rm nuc}$ or q that are due to a temperature perturbation, and our foregoing arguments remain valid.

6.3 Dynamical stability

Dynamical stability is related to motions of mass parcels in the star, that is, to macroscopic motions; on the microscopic scale, the gas particles are always in random, local motion. In hydrostatic equilibrium, no macroscopic motions occur; more precisely, they occur imperceptibly slowly. In order to test the stability of this equilibrium, we have to consider the response to small perturbations of the balance between the gravitational attraction and the outward force exerted by the pressure gradient. Since we deal with a spherically symmetric configuration, we shall consider radial perturbations: compression or expansion. The basic question is whether a temporary contraction will result in expansion toward the original state or in further contraction, escalating in a runaway.

A rigorous treatment of dynamical perturbations within a star is far from simple. But in order to illustrate the basic principles involved, a highly simplified example should suffice. Consider a gaseous sphere of mass M, in hydrostatic equilibrium. The pressure at any point $r(m)$ is equal to the weight per unit area of the layers between m and M, as obtained by integrating the equation of hydrostatic equilibrium (5.1), and taking $P(M) = 0$:

$$P = \int_m^M \frac{Gm}{4\pi r^4} \, dm. \tag{6.14}$$

The density at $r(m)$ is given by equation (5.2):

$$\rho = \frac{1}{4\pi r^2} \frac{dm}{dr}. \tag{6.15}$$

Consider now a small, uniform, radial compression, so that the new radii are everywhere obtained from the original ones by a small perturbation:

$$r' = r - \varepsilon r = r(1 - \varepsilon). \tag{6.16}$$

If $\varepsilon \ll 1$, the binomial approximation

$$(1 \pm \varepsilon)^n \approx 1 \pm n\varepsilon \tag{6.17}$$

can be used. The new densities will be

$$\rho' = \frac{1}{4\pi r^2 (1-\varepsilon)^2} \frac{dm}{dr} \frac{dr}{dr'} = \frac{\rho}{(1-\varepsilon)^3} \approx \rho(1 + 3\varepsilon). \tag{6.18}$$

Assuming furthermore the contraction to be adiabatic (and neglecting radiation pressure), we find that the new gas pressure will relate to the initial one as $(\rho'/\rho)^{\gamma_a}$, where γ_a is the adiabatic exponent (introduced in Section 3.6):

$$P'_{\text{gas}} = P(1 + 3\varepsilon)^{\gamma_a} \approx P(1 + 3\varepsilon\gamma_a). \tag{6.19}$$

Similarly, by equation (6.14), the new hydrostatic pressure will relate to the initial one as

$$P'_{\text{h}} = \int_m^M \frac{Gm}{4\pi r^4 (1-\varepsilon)^4} \, dm \approx P(1 + 4\varepsilon). \tag{6.20}$$

It is to be expected that after this perturbation the gaseous sphere will no longer be in hydrostatic equilibrium, that is, $P'_{\text{gas}} \neq P'_{\text{h}}$. The condition required for restoring equilibrium is in our case

$$P'_{\text{gas}} > P'_{\text{h}}, \tag{6.21}$$

so as to cause the sphere to expand back to its original state. Substituting equations (6.19) and (6.20) into equation (6.21), we thus require

$$P(1 + 3\gamma_a\varepsilon) > P(1 + 4\varepsilon). \tag{6.22}$$

Hence the condition for *stable* hydrostatic equilibrium, or in other words, the condition for *dynamical stability*, is

$$\gamma_a > \frac{4}{3}. \tag{6.23}$$

The same result obtains in the case of expansion, when $\varepsilon < 0$ and condition (6.21) is reversed.

It can be shown rigorously that a star in which $\gamma_a > 4/3$ *everywhere* is dynamically stable (and neutrally stable, if $\gamma_a = 4/3$ everywhere). The case in which $\gamma_a < 4/3$ *somewhere* requires further examination. Global dynamical instability is obtained if the integral $\int (\gamma_a - 4/3)\frac{P}{\rho}dm$ over the entire star is negative. Thus, if $\gamma_a < 4/3$ in a sufficiently large core, where P/ρ is high, the star will become unstable. If, however, $\gamma_a < 4/3$ in the outer layers, where P/ρ is small, the star as a whole need not become unstable.

6.4 Cases of dynamical instability

The question we have to ask is, What are the stellar configurations that may lead to violation of the stability criterion, that is, to $\gamma_a \leq 4/3$? We have already encountered such cases in Section 3.6.

Relativistic-degenerate electron gas

For a relativistic-degenerate electron gas, γ_a tends to 4/3. The instability expected in the limit $\gamma_a = 4/3$ results in this case in the Chandrasekhar limiting mass (derived in Section 5.4): the degeneracy pressure can sustain the gravitational attraction only if the stellar mass is smaller than this limit. For a higher mass, contraction will end in collapse.

Dominant radiation pressure

The second case for which γ_a tends to 4/3 (as shown in Section 3.6) is that of a dominant radiation pressure, or in terms of the parameter β – introduced in Section 3.1 – $\beta \to 0$. In the limit $\beta = 1$ (ideal gas without radiation), $\gamma_a = 5/3$, and hence an ideal gas would be dynamically stable under its own gravitational field. As β decreases, γ_a decreases as well, tending to 4/3 in the limit $\beta = 0$ (pure photon gas). Another way of showing that a radiation pressure dominated gas tends to be dynamically unstable is by using the virial theorem. For pure radiation $P/\rho = u_{\rm rad}/3$ and hence by equation (2.23)

$$-\Omega = 3\int_0^M \frac{P}{\rho}dm = U_{\rm rad}, \tag{6.24}$$

meaning that the total energy of a star $E = \Omega + U$ vanishes; that is, the star becomes unbound. We see, therefore, that the consequences of dynamical instability may differ.

Ionization-type processes

Dynamical instability, or $\gamma_a \leq 4/3$, is also prone to occur in any system of particles in which the number of particles is not conserved, but changes with changing physical conditions. Ionization (Section 3.6) provides a typical example: a single atom may produce two particles, an ion and an electron, by absorbing the right amount of energy from a collision with another particle or with a photon. At the same time the reverse reaction – *recombination* – occurs, which tends to diminish the number of particles. When the system is compressed, recombination is enhanced, whereas if the volume is increased, ionization is favoured. Therefore, the number of particles changes in inverse proportion to the density. The following simple argument is meant to provide an intuitive explanation to the effect of this property on the value of γ_a. Consider two systems of particles of volume V and pressure P: in one the number of particles N is conserved; in the other it may change due to ionization-type reactions. We assume an ideal gas and recall that the pressure is proportional to the *number* of particles (regardless of their nature) and inversely proportional to the *volume*. Suppose now that the volume is slightly compressed to $V' < V$. In the first system the pressure would obviously increase, since $N/V' > N/V$. In the second system, however, N would change as well, say, to $N' < N$. Hence the new ratio N'/V' would be smaller than in the first system, $N'/V' < N/V'$. Consequently, the pressure would increase to a lesser extent, meaning that the dependence of the pressure on volume (and hence, density) is *weaker* in the second system. This should translate into a smaller value of γ_a, possibly smaller than 4/3. For a pure, singly ionized gas, for example, according to equation (3.60), $\gamma_a < 4/3$ between 18% and 82% ionization (for $kT \approx \chi$). Hence only an almost entirely neutral or a completely ionized gas would be dynamically stable.

Since in stellar interiors temperatures are sufficiently high to ensure total ionization – at least for the major components, hydrogen and helium – ionization in itself is of no great consequence regarding the global stability of stars ($\gamma_a < 4/3$ in restricted zones, where P/ρ is small). We have encountered, however, two other ionization-type processes, which occur at high temperatures: iron photodisintegration (Section 4.10) and pair production (Section 4.9). We shall see in the next chapter that both these processes drastically affect the course of stellar evolution.

6.5 Convection

We have seen in Section 5.5 that the radiative energy flux through a star in hydrostatic equilibrium is limited by the requirement

$$\kappa F < 4\pi c G m,$$

which may be violated in cases of intense nuclear burning, when $F = \int q\, dm$

6.5 Convection

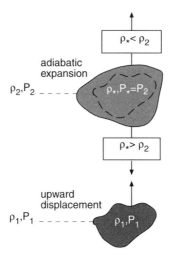

Figure 6.2 Illustration of Schwarzschild's criterion for stability against convection by a mass element dm moving radially from point 1 to point 2 within a star.

becomes exceedingly high, or when the opacity is very high. By equation (5.3), a high flux or a high opacity leads to a steep temperature gradient. However, the temperature gradient may only increase up to a limit beyond which convection occurs, involving cyclic macroscopic mass motion (but not a net mass flux) that carries the bulk of the energy flux. When the total flux satisfies equation (5.4), (convective) thermal equilibrium is achieved. Thus convection may be regarded as a type of dynamical instability, although it does not have disruptive consequences. In fact, in spite of being a dynamic process, convection affects the structure of a star only as an effective heat carrier and as a mixing mechanism. The condition for the onset of convection (or the limiting temperature gradient) is determined by a simple criterion, as was shown by Karl Schwarzschild in 1906.

The Schwarzschild criterion for stability against convection derives from the following argument: Consider a mass element Δm at some point within a star, as shown in Figure 6.2. Denoting this point by 1; let the local values of density and pressure at 1 be ρ_1 and P_1, respectively. Suppose the element moves a small distance outward in the radial direction to point 2, where the density is ρ_2 and the pressure is P_2. Since the pressure in a star decreases outward, $P_2 < P_1$; that is, the surrounding pressure at point 2 will be lower than the pressure within the mass element. The element will therefore expand until the internal and external pressures are in balance. In view of the great difference between dynamical and thermal timescales, it is reasonable to assume that no heat exchange with the environment occurs while the mass element expands. Hence the element undergoes an adiabatic change leading to a final density ρ_*, which is not necessarily equal to the density of its surroundings. If $\rho_* > \rho_2$, the mass element will descend

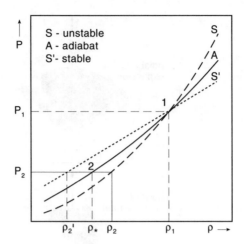

Figure 6.3 Schematic density-pressure diagram, leading to the mathematical formulation of Schwarzschild's criterion for stability against convection.

back toward its initial position. We regard such a situation as *stable*, for any mass motion that may accidentally arise will be damped. If, on the other hand, $\rho_* < \rho_2$, the element will continue its upward motion (by the Archimedes law). In this case, the system is unstable against convection; that is, convective motion is prone to develop. The extent of the convectively unstable region may be found by applying the same criterion for increasingly more distant points. It is possible that a star be fully convective, all the way from the centre to the surface.

To obtain a mathematical formulation of the convective stability condition, we resort to the (ρ, P) diagram of Figure 6.3, where the starting point $1 - (\rho_1, P_1)$ – is marked. As shown in Section 3.6, the dependence of pressure on density in an adiabatic process is given by $P = K_a \rho^{\gamma_a}$. The curve labelled A represents the adiabatic (P, ρ) relation passing through point 1, obtained from the physical characteristics of the gas at that point. The curves labelled S and S' represent hypothetical stellar configurations: possible variations of the pressure with density in the star in the neighbourhood of point 1. The slope of S is steeper and that of S' is shallower than the slope of A. The horizontal line $P = P_2$ intersects each of the curves A, S, and S': the intersection with A corresponds to the density ρ_* within the mass element, while the intersections with S and S' correspond to the density of the surroundings in each case. If the stellar configuration is described by S, then $\rho_2 > \rho_*$, meaning instability, whereas if S' describes the stellar configuration, $\rho_2' < \rho_*$, indicating stability against convection. In conclusion, the condition for stability is

$$\left(\frac{dP}{d\rho}\right)_{\text{star}} < \left(\frac{dP}{d\rho}\right)_a, \qquad (6.25)$$

and multiplying both sides by ρ/P, we have

$$\gamma \equiv \frac{\rho}{P}\left(\frac{dP}{d\rho}\right)_{\text{star}} < \gamma_a. \tag{6.26}$$

It is noteworthy that the general validity of this simple criterion was proved by rigorous mathematical methods only six decades after it came into use, in 1967, by Shmuel Kaniel and Attay Kovetz.

For an ideal gas and negligible radiation pressure, the pressure is proportional both to temperature and to density, whence

$$\frac{dP}{P} = \frac{d\rho}{\rho} + \frac{dT}{T} \tag{6.27}$$

for a given composition. Combining equations (6.27) and (6.26), we obtain the condition for convective stability in the form

$$\frac{P}{T}\left(\frac{dT}{dP}\right)_{\text{star}} < \frac{\gamma_a - 1}{\gamma_a}, \tag{6.28}$$

which may also be written as

$$\left|\frac{dT}{dr}\right|_{\text{star}} < \left(\frac{\gamma_a - 1}{\gamma_a}\right)\frac{T}{P}\left|\frac{dP}{dr}\right|_{\text{star}}, \tag{6.29}$$

recalling that the temperature and the pressure gradients are negative. We have thus obtained the upper limit for the magnitude of the temperature gradient allowed before convection sets in.

6.6 Cases of convective instability

The criterion for convective stability that we have just derived is very general; it may be equally applied to stellar interiors and, for example, to the Earth's atmosphere. But can we be more specific about the conditions that may lead to convective instability in stars? In particular, how is restriction (5.34) connected to the criterion of convective stability?

We have seen in Section 3.6 that during ionization the adiabatic exponent is lowered. Therefore, in regions of the star where the gas is partially ionized, the condition for convective stability is more difficult to satisfy. At the same time these regions may become dynamically unstable, if $\gamma_a < 4/3$.

Condition (6.28) may be generalized to include the effect of radiation pressure, in which case $\gamma_a = \gamma_a(\beta)$, but the adiabatic exponents that appear in conditions (6.26) and (6.28) are different functions of β. Both exponents tend to 5/3 for $\beta \to 1$ and to 4/3 for $\beta \to 0$.

Exercise 6.2: Following the procedure of Section 3.6, derive the adiabatic exponents in conditions (6.26) and (6.28) for an ideal gas and radiation, as functions of β. Calculate their values for $\beta = 0$, $\frac{1}{2}$, and 1.

If we now use the radiative diffusion equation (5.3) for the temperature gradient and the hydrostatic equation (5.1) for the pressure gradient, we obtain the condition for convective stability (6.29) in the form

$$\kappa F < 4\pi c G m \left[4 \left(\frac{\gamma_a - 1}{\gamma_a} \right) (1 - \beta) \right]. \tag{6.30}$$

This is similar to condition (5.34), which imposed an upper limit on the product κF, above which radiative equilibrium could no longer hold. We note that condition (6.30) is stronger, since the term in square brackets on the right-hand side is smaller than unity. Therefore convection arises *before* the upper limit for κF [condition (5.34)] is reached. The two conditions converge as β tends to zero.

In the case of ionization, the high opacity and low adiabatic exponent combine to induce convection. This effect is particularly important in the outer regions of stars, where temperatures are sufficiently low for helium and hydrogen to be only partially ionized. In stellar interiors, especially in zones of high temperature where the opacity is constant, the dominant factor that may induce convection is a high energy flux. Such a flux is expected to be obtained from intense nuclear burning. Assuming that the nuclear energy generation rate may be expressed as a power law of the form (5.7), $q = q_0 \rho^m T^n$ with $n \gg m$, it should be possible to translate the condition on the intensity of nuclear burning into a limiting value for n. This is by no means a simple task; it cannot be accomplished analytically but requires solution of the stellar structure equations. The question of a limiting value for n has already been addressed in the 1930s; for example, using relatively simple models (more elaborate variants of the models described in Section 5.7), Cowling obtained the following conditions: for a constant opacity, no convectively stable configuration is possible if n exceeds a number lying between 3 and 4, while for a Kramers opacity law (Section 3.7), no such solution is possible for n in excess of about 8. In the early 1950s the problem was pursued and elaborated by Roger Tayler, with similar results. Generally, a high temperature sensitivity of the energy generation rate is bound to trigger convection.

Condition (6.30) indicates that convection is more likely to occur when besides κF being high, β is not too far from unity, meaning that gas pressure is dominant. We have seen in Section 5.6 that, based on the simple *standard model*, β is strongly related to the stellar mass, increasing with decreasing M. Hence, we should not be surprised to find that convection is dominant in low-mass stars burning nuclear fuel. When low-mass stars are sufficiently cold and dense for degeneracy pressure to dominate, stability against convection is regained. This is for two reasons: first, degenerate matter is highly conductive, that is, its effective opacity is very low,

6.6 Cases of convective instability

and secondly, no stable nuclear burning is possible under degenerate conditions (Section 6.2 above), implying that such stars must be inert. In conclusion, we should not expect convection to develop in the interiors of white dwarfs.

When convective energy transport takes place, equation (5.3) is no longer valid in the sense that the flux appearing on the right-hand side is the radiation flux, which now differs from the total flux F [of equation (5.4)], amounting to only a small, unknown fraction of F. Hence equation (5.3) must be supplemented or replaced by another equation that takes account of convection. Since convective motions are clearly not entirely radial, there are only approximate ways of estimating the convective flux for spherical, one-dimensional stellar models. These methods are better suited to more advanced texts. For our purposes, it is important to note that within convective zones in stellar interiors the departure from adiabaticity is very small ($\gamma \approx \gamma_a$). This was first shown by Ludwig Biermann in the 1930s, essentially by the following argument.

Consider a convective shell of mass m_c and thickness r_c located at a distance r (mass m) from the centre. Let F be the energy crossing the inner boundary of the convective shell per unit time. In the case of convective transfer, the energy is first absorbed and then transmitted by turbulent mass motions. Thus a rising mass element must acquire a surplus temperature δT over the temperature T of its surroundings in order to transport heat, and $\delta T/T$ is a measure of the superadiabaticity. Since the mass element is in dynamical equilibrium with its surroundings, $\delta P = 0$. Thus, from equation (6.27), its density deficit is related to the temperature surplus by $|\delta \rho / \rho| = |\delta T / T|$. The time required for the mass element to convect through the distance r_c and transfer heat is r_c/v_c, where v_c is its average velocity. Hence, on the whole, an amount of energy $F(r_c/v_c)$ is absorbed in the shell before it is carried away through the outer boundary. The relative rise in temperature is then given by

$$\frac{\delta T}{T} \sim \frac{F(r_c/v_c)}{u\, m_c}, \tag{6.31}$$

where u is the energy per unit mass (see Section 3.5). The convective velocity may be roughly estimated by $v_c \sim \sqrt{g' r_c}$, where g' is the acceleration, which, by the Archimedes buoyancy principle, is the local gravitational acceleration, reduced by the factor $\delta \rho / \rho$. Hence

$$v_c \sim \sqrt{\frac{Gm}{r^2}\left|\frac{\delta \rho}{\rho}\right| r_c} \sim \sqrt{\frac{Gm}{r^2}\frac{\delta T}{T} r_c}. \tag{6.32}$$

Combining relations (6.31) and (6.32), we have

$$\left(\frac{\delta T}{T}\right)^{3/2} \sim \frac{Fr}{u\, m_c \sqrt{Gm/r_c}}. \tag{6.33}$$

To obtain an order of magnitude estimate, we replace F by the stellar luminosity L, m and m_c by the stellar mass M, uM by U, and r and r_c by the stellar radius R, which yields

$$\left(\frac{\delta T}{T}\right)^{3/2} \sim \frac{L}{U}\frac{1}{\sqrt{GM/R^3}}. \tag{6.34}$$

We recognize the first term of the product on the right-hand side as the reciprocal of the thermal (Kelvin–Helmholtz) timescale [equation (2.60)], of the order of 10^{15} s, and the second as the dynamical timescale [equation (2.56)], of the order of 10^3 s. In conclusion,

$$\frac{\delta T}{T} \sim \left(\frac{\tau_{\rm dyn}}{\tau_{\rm th}}\right)^{2/3} \sim 10^{-8}, \tag{6.35}$$

which clearly proves our claim. Thus, to a very good approximation, the temperature gradient may be assumed to be adiabatic in deep convective zones [instead of equation (5.3)], and hence such zones are polytropic. This assumption is not valid close to the stellar surface, where $uM \ll U$.

> **Exercise 6.3:** Assuming that a star of uniform κ (opacity) and β has a convective core, and no nuclear energy generation outside the core, show that the mass fraction of this core is given by $\frac{\gamma_a}{4(\gamma_a - 1)}$.

Now, since dynamical stability requires $\gamma_a > 4/3$, a star in hydrostatic equilibrium must satisfy

$$\frac{4}{3} < \gamma_a \leq \frac{5}{3}, \tag{6.36}$$

which means that if the configuration of a star is to be approximately described by a polytrope (in which case γ and γ_a are identical), the index n may only vary between 1.5 and 3.

6.7 Conclusion

To summarize the question of stability of equilibrium in a star (whether radiative or convective) raised at the beginning of this chapter, let us say that stability depends on the ability of the gas – particles and photons – at any given point to sustain the weight of the overlying layers by means of the pressure it exerts, so as to maintain exact balance despite possible perturbations. To this end, the pressure must depend strongly enough both on temperature and on density. Sensitivity to temperature is required in order to prevent thermonuclear runaways, and to density – so as to avoid collapse. There are additional cases of instability in stars, essentially resulting from violation of this principle, but since they occur in the course of evolution when special interior configurations develop, we shall deal with them as they arise.

7

The evolution of stars – a schematic picture

Having answered the two basic questions posed at the the end of Chapter 2, our present task is to combine the knowledge acquired so far into a general picture of the evolution of stars. We recall that the timescale of stellar evolution is set by the (slow) rate of consumption of the nuclear fuel. Now, the rate of nuclear burning increases with density and rises steeply with temperature, and the structure equations of a star show that both the temperature and the density decrease from the centre outward. We may therefore conclude that the evolution of a star will be led by the central region (the *core*), with the outer parts lagging behind it. Changes in composition first occur in the core, and as the core is gradually depleted of each nuclear fuel, the evolution of the star progresses.

Thus insight may be gained into the evolutionary course of a star by considering the changes that occur at its centre. To obtain a simplified picture of stellar evolution, we shall characterize a star by its central conditions and follow the change of these conditions with time. We have seen that besides the composition, the temperature and density are the only properties required in order to determine any other physical quantity. If we denote the central temperature by T_c and the central density by ρ_c, the state of a star is defined at any given time t by a pair of values: $T_c(t)$ and $\rho_c(t)$. Consider now a diagram whose axes are temperature and density. The pair $[T_c(t), \rho_c(t)]$ corresponds to a point in such a diagram, and the evolution of a star is therefore described by a series of points, $[T_c(t_1), \rho_c(t_1)], [T_c(t_2), \rho_c(t_2)], [T_c(t_3), \rho_c(t_3)], [T_c(t_4), \rho_c(t_4)], \ldots$, for times $t_1 < t_2 < t_3 < t_4, \ldots$, which forms a *parametric* line $[T_c(t), \rho_c(t)]$. Since the only property that distinguishes the evolutionary track of a star from that of any other star of the same composition is its mass, we may expect to obtain different lines in the (T, ρ) plane for different masses.

Note: A study of the *late stages* of stellar evolution, based on homogeneous and isentropic (uniform entropy or adiabatic structure) models, was performed by Gideon Rakavy and Giora Shaviv in 1968. The progress in time was simulated by decreasing the entropy s, and thus parametric lines $[T_c(s), \rho_c(s)]$ were obtained for different values of the stellar mass. A beautiful general picture of the end states of stars emerged, for which a qualitative explanation (and, in a sense, validation) was offered a year later by Kovetz. The following discussion, which will eventually lead to a more comprehensive picture arising from very simple arguments, was inspired by these works.

All the processes that are bound to occur in a star have characteristic temperature and density ranges, and hence different combinations of temperature and density will determine the prevailing state of the stellar material and the dominant physical processes that should be expected to occur. Thus the (T, ρ) plane may be divided into zones, representing different physical states or processes. Our first step will be to get acquainted with the terrain through which the evolution paths of the stellar centre are winding; the second step will be to identify the track corresponding to each stellar mass; finally, by following each track through this terrain, we shall be able to trace the chain of processes that make up the evolution of a star.

7.1 Characterization of the (log T, log ρ) plane

The (T, ρ) plane will be divided into zones dominated by different equations of state and different nuclear processes. Of particular interest will be those regions where the conditions are bound to lead to dynamical instability. As the ranges of density and temperature typical of stellar interiors span many orders of magnitude, logarithmic scales will be used for both.

Zones of the equation of state

The following arguments are based on the material of Chapter 3 and lead to Figure 7.1. The most common state of the ionized stellar gas is that of an ideal gas for both components: ions and electrons. Hence the common equation of state is of the form

$$P = \frac{\mathcal{R}}{\mu}\rho T = K_0 \rho T, \tag{7.1}$$

where K_0 is a constant [see equation (3.28)]. At high densities and relatively low temperatures, the electrons become degenerate, and since their contribution to the pressure is dominant, the equation of state may then be approximated by

$$P = K_1 \rho^{5/3} \tag{7.2}$$

[see equation (3.35)], which replaces equation (7.1). The transition from one state to the other is, of course, gradual with the change in density and temperature, but

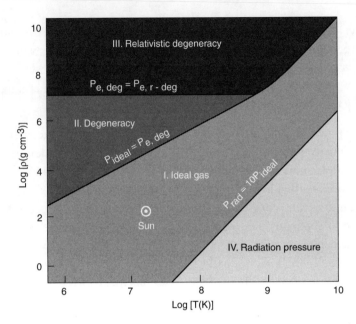

Figure 7.1 Mapping of the temperature-density diagram according to the equation of state.

an approximate boundary may be traced in the ($\log T$, $\log \rho$) plane on one side of which the effect of degeneracy is clearly important, while on the other side an ideal gas law prevails. This boundary may be defined by the requirement that the pressure obtained from equation (7.1) be equal to that obtained from equation (7.2),

$$\log \rho = 1.5 \log T + \text{constant}, \tag{7.3}$$

which is a straight line with a slope of 1.5, as shown in Figure 7.1. The electron degeneracy zone, labelled II in Figure 7.1, where

$$K_1 \rho^{5/3} > K_0 \rho T,$$

lies above (to the left of) this line. The ideal gas zone, labelled I, lies below it.

For still higher densities, when relativistic effects play an important role, the equation of state changes to the form

$$P = K_2 \rho^{4/3} \tag{7.4}$$

[see equation (3.38)]. The boundary between the ideal gas zone and the relativistic-degeneracy zone may be obtained, as before, from the requirement

$$K_2 \rho^{4/3} = K_0 \rho T,$$

which defines a straight line

$$\log \rho = 3 \log T + \text{constant} \qquad (7.5)$$

with a slope of 3. Thus the boundary between the ideal gas zone and the electron degeneracy zone changes slope, becoming steeper as the density increases.

Within the electron degeneracy region, the transition from nonrelativistic to relativistic degeneracy occurs when the rise in pressure with increasing density becomes constrained by the limiting velocity c. Hence relativistic degeneracy should be considered when

$$K_1 \rho^{5/3} \gg K_2 \rho^{4/3},$$

or

$$\rho \gg \left(\frac{K_2}{K_1}\right)^3, \qquad (7.6)$$

that is, above a high pressure level [a horizontal line in the (log T, log ρ) plane]. This is roughly indicated in Figure 7.1, where the relativistic-degeneracy zone is labelled III.

In zone I radiation pressure has been neglected. Its contribution to the total pressure becomes important, however, at high temperatures and low densities (the lower right corner of the diagram) and should be *added* to that of the gas. Eventually, radiation pressure would become dominant, with the equation of state changing to

$$P = \tfrac{1}{3} a T^4 \qquad (7.7)$$

[see equation (3.40)]. Taking the gas pressure to be negligible for, say, $P_{\text{rad}} = 10 P_{\text{gas}}$, we obtain an approximate boundary for the zone of dominance of radiation pressure (labelled IV in Figure 7.1) in the form

$$\log \rho = 3 \log T + \text{constant} \qquad (7.8)$$

again a straight line of slope 3 [with a different constant than in equation (7.5), of course].

Zones of nuclear burning

The following arguments are based on the material of Chapter 4. A nuclear burning process of any kind becomes important in a star whenever the rate of energy release by this process constitutes a significant fraction of the rate at which energy

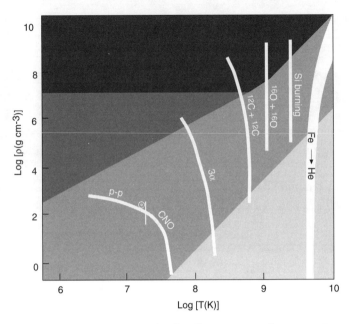

Figure 7.2 Mapping of the temperature-density diagram according to nuclear processes.

is radiated away, that is, the stellar luminosity. Although stellar luminosities vary within a wide range, the variation in the conditions prevailing in burning zones is quite restricted, due to the high sensitivity of nuclear reaction rates to temperature. Hence a narrow *threshold* may be defined for each nuclear process that takes place in stars. On one side of the threshold the rate of nuclear burning may be assumed negligible, and on the other side, considerable. The threshold for each process constitutes a line in the $(\log T, \log \rho)$ plane, defined by the requirement that the rate of nuclear energy generation q exceed a certain prescribed limit $q_{\min}$, say, 0.1 J kg^{-1} s^{-1} (10^3 erg g^{-1} s^{-1}). Since for each process, q may be approximated by a power law of the form

$$q = q_0 \rho^m T^n, \tag{7.9}$$

the threshold given by $q = q_{\min}$ is

$$\log \rho = -\frac{n}{m} \log T + \frac{1}{m} \log\left(\frac{q_{\min}}{q_0}\right). \tag{7.10}$$

The exoergic transformation of hydrogen into the iron group elements comprises five major stages: hydrogen burning into helium either by the p–p chain or by the CNO cycle, helium burning into carbon by the 3α reaction, carbon burning, oxygen burning, and silicon burning. The five corresponding thresholds are plotted in Figure 7.2. In most of the cases $m = 1$ and $n \gg 1$, and hence the (negative) slope in equation (7.10) is so steep that the thresholds are almost vertical lines. Strictly, the threshold defined by equation (7.10) should be a straight line; in reality, the values of the powers in equation (7.9) change slightly for different temperature ranges; this is the reason why the lines in Figure 7.2 are not perfectly straight. For hydrogen burning, the slope is milder at low temperatures, corresponding to the p–p chain ($n \approx 4$), and becomes steeper at higher temperatures, where the CNO cycle ($n \approx 16$) becomes dominant.

Nucleosynthesis by energy releasing fusion of lighter elements into heavier ones ends with iron. Iron nuclei heated to very high temperatures are disintegrated by energetic photons into helium nuclei. This energy absorbing process reaches equilibrium (called, as in the case of silicon burning, *nuclear statistical equilibrium*), with the relative abundance of iron to helium nuclei determined by the values of temperature and density. A threshold may be defined for the process of iron photodisintegration, as a strip in the ($\log T$, $\log \rho$) plane, by the requirement that the number of helium and iron nuclei be approximately equal. This threshold is shown in Figure 7.2.

Zones of instability

The following arguments are based on the material of Chapter 6. The condition of dynamical stability is $\gamma_a > 4/3$ (Section 6.3). We thus expect stellar configurations to become *dynamically unstable* in those regions of the ($\log T$, $\log \rho$) plane where γ_a is reduced to $4/3$ or less. Such regions are the far extremes of the relativistic degeneracy zone III and of the radiation pressure dominated zone IV, where γ_a tends asymptotically to $4/3$. Another is the iron photodisintegration zone, where $\gamma_a < 4/3$. As we are dealing with the centre of stars, restricted regions of instability caused by the ionization of hydrogen and helium lie outside the ranges of temperature and density that we consider. Pair production, which is an ionization-type process as well, defines an additional unstable zone, with $\gamma_a < 4/3$, as shown in Figure 7.3. With all these unstable zones marked, the stable part of the ($\log T$, $\log \rho$) plane becomes completely bounded on two sides: at high densities and at high temperatures. Hence severe constraints are imposed on the possible evolutionary tracks of stars. We finally recall that nuclear burning is *thermally unstable* in degenerate gases, whether relativistic or not. Hence the nuclear burning thresholds of Figure 7.2 have been discontinued after crossing the boundary into the degeneracy zone II.

7.2 The evolutionary path of the central point of a star in the ($\log T$, $\log \rho$) plane

Having become acquainted with the ($\log T$, $\log \rho$) plane, the question we now ask is whether the centre of a star of given mass M may assume any combination of temperature and density values, that is, may be found anywhere in this plane,

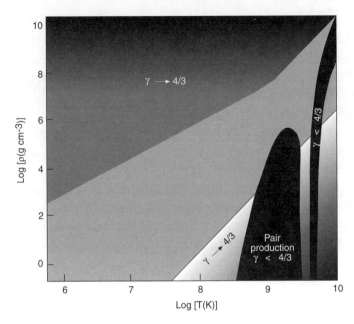

Figure 7.3 Outline of the stable and unstable zones in the temperature-density diagram.

or whether these values are in some way constrained by M. We now regard the ($\log T$, $\log \rho$) plane as a ($\log T_c$, $\log \rho_c$) plane, referring to the stellar centre. Assuming a polytropic configuration [equation (5.10)] for a star in hydrostatic equilibrium, the central density is related to the central pressure by equation (5.28),

$$P_c = (4\pi)^{1/3} B_n G M^{2/3} \rho_c^{4/3}. \tag{7.11}$$

This relation is only weakly dependent on the polytropic index n, especially for stable configurations, for which n varies between 1.5 and 3 (cf. Section 6.6) and the coefficient B_n between 0.157 and 0.206 (cf. Table 5.1), and it is independent of K. It is valid whether K is determined by processes on the microscopic scale, such as electron degeneracy (Section 3.3), or on the macroscopic scale, such as convection (Section 6.6). Although a star in hydrostatic equilibrium is *not* a perfect polytrope (even if its composition is homogeneous), relation (7.11) provides a good approximation to hydrostatic equilibrium for *any* configuration. Note that simply by dimensional analysis of the hydrostatic equation, the central pressure must be proportional to $GM^{2/3}\bar\rho^{4/3}$.

In addition, the central pressure is related to the central density and temperature by the equation of state. Within the different zones of the ($\log T_c$, $\log \rho_c$) plane we have different equations of state. Combining each of them with equation (7.11), we may eliminate P_c, to obtain a relation between ρ_c and T_c.

Consider a star of mass M, whose central point is found in the ideal gas zone I, where equation (7.1) holds. The relation between T_c and ρ_c in this case is of

7.2 The evolutionary path of the central point of a star

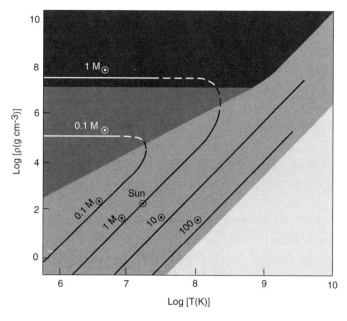

Figure 7.4 Relation of central density to central temperature for stars of different masses within the stable ideal gas and degenerate gas zones.

the form

$$\rho_c = \frac{K_0^3}{4\pi B_n^3 G^3} \frac{T_c^3}{M^2}, \tag{7.12}$$

meaning that for a star of given mass, the central density varies as the central temperature cubed. For stars of different masses but the same central temperature, the central density decreases as the mass squared increases. On logarithmic scales, relation (7.12) becomes a straight line with a slope of 3. Thus different masses define different *parallel* lines, which intersect the temperature axis at intervals proportional to $\log M$. The lines corresponding to $M = 0.1, 1, 10,$ and $100 M_\odot$ are plotted in Figure 7.4; these masses being successive powers of $10 M_\odot$, the intervals between the lines are equal.

If at the centre of a star the electrons are strongly but nonrelativistically degenerate, the central point is found in zone II and equation (7.2) holds. Substituting P_c from equation (7.11) with $n = 1.5$, we obtain

$$\rho_c = 4\pi \left(\frac{B_{1.5} G}{K_1}\right)^3 M^2, \tag{7.13}$$

which replaces the ideal gas relation (7.12). Here ρ_c is independent of T_c, and the corresponding line in the $(\log T_c, \log \rho_c)$ plane is horizontal at a height that increases with mass M, as plotted in Figure 7.4. Strictly, from equation (7.13) the central density should vary as the mass squared, but relativistic effects increase the power. Zones I and II are the only stable regions in the $(\log T, \log \rho)$ plane and hence we need not consider the others.

For relatively low masses, relations (7.12) and (7.13) will merge at the boundary between zones I and II, as shown by the dashed segments in Figure 7.4, resulting in a continuous bending path characteristic of each mass. We have seen in Section 5.3 that the density of degenerate stars tends to infinity as the stellar mass approaches the critical Chandrasekhar limit M_{Ch} – the highest mass that can be sustained in hydrostatic equilibrium by electron-degeneracy pressure. Thus the paths corresponding to increasing masses will bend at higher and higher density values in the $(\log T, \log \rho)$ plane, deeper into the region of relativistic degeneracy. It is easy to see that the limiting case will be represented by a straight line, which will also mark the division between paths that bend into the degeneracy zone II and those which remain in zone I. We recall that the boundary between the ideal gas zone and the degenerate electron gas zone, close to its relativistic part, has a slope of 3. Since the $(\log T_c, \log \rho_c)$ curves in the ideal gas zone have a slope of 3 as well, there exists a value of M that coincides with the relativistic-degeneracy boundary. This mass is M_{Ch}, which was obtained by equating the right-hand sides of equations (7.4) and (7.11), while the boundary between zones I and III was obtained by equating the right-hand sides of equations (7.11) and (7.1). Hence the boundary between zones I and II merges with the path corresponding to M_{Ch} in the $(\log T_c, \log \rho_c)$ plane.

In conclusion, a star of fixed mass has its own distinct track in the $(\log T_c, \log \rho_c)$ plane, which we shall refer to in the following text as Ψ_M. There are two characteristic shapes of Ψ_M: straight lines for $M > M_{Ch}$ and knee-shaped ones for $M < M_{Ch}$. In general terms, we may understand the relationship between tracks corresponding to different masses as follows. With increasing stellar mass, the gravitational pull toward the centre becomes stronger. Hence a higher pressure is required to counterbalance gravity. This may be achieved in an ideal gas by a higher density or a higher temperature. A higher density implies, however, smaller distances between material particles, which further enhance the gravitational field. In fact, since the hydrostatic pressure is proportional to a *higher* power of the density than is the gas pressure (4/3 as compared to 1), a higher density would only worsen the imbalance. Thus a *lower* density or a *higher* temperature is required for equilibrium, if the stellar mass is increased. In the case of a degenerate electron gas, the temperature plays a far less important role. But now the hydrostatic pressure is proportional to a *lower* power of the density than is the gas pressure (4/3 as compared to $\sim 5/3$), so that a higher density is needed for equilibrium in a more massive star.

The question we now have to answer is, Where does the evolutionary course of a star lead the central point along a track?

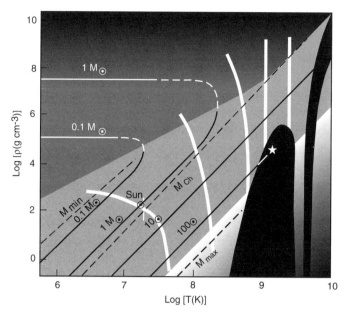

Figure 7.5 Schematic illustration of the evolution of stars according to their central temperature-density tracks.

7.3 The evolution of a star, as viewed from its centre

Combining Figures 7.1 to 7.4 into one picture, we obtain a full, albeit schematic, view of stellar evolution, as shown in Figure 7.5. We may now choose a mass M, identify its path (marked in Figure 7.5 by the value of M), and follow the journey of the ($\log T_c$, $\log \rho_c$) point along it, to discover what it encounters on its way. Stars form in gaseous clouds, where densities and temperatures are much lower than those prevailing in stellar interiors; therefore, the starting point is on the lower part of the path. At the beginning, a star radiates energy without an internal energy source, which means that it contracts and heats up (as we have seen in Chapter 2 and again in Chapter 6). Hence in the ($\log T_c$, $\log \rho_c$) plane the central point ascends along Ψ_M – which we recall to be a straight line of slope 3 – toward higher temperatures and densities. Eventually, it will cross the first nuclear burning threshold. At this point in the evolution of a star hydrogen is ignited at the centre and the star adjusts into thermal equilibrium with L_{nuc} and L in balance. The journey of the central point along Ψ_M comes to a very long pause. We note that for low masses Ψ_M crosses the threshold on the upper part, corresponding to the p–p chain, whereas for high masses the threshold is crossed on the lower part, corresponding to the CNO cycle. We should therefore expect stars to burn hydrogen by different processes according to their masses.

It was shown in the previous section that the boundary between the ideal gas zone and the radiation pressure instability zone has a slope of 3 (regardless of the criterion adopted for its definition), the same as the slopes of the Ψ_M curves. Hence as the mass increases, Ψ_M inevitably approaches this boundary. This means that in massive stars radiation pressure becomes progressively more important and eventually dominates gas pressure. Since a star dominated by radiation pressure is dynamically unstable (becoming unbound), an upper limit thus emerges for the stellar mass, roughly near (or somewhat above) $100M_\odot$, as marked by the curve $\Psi_{M_{max}}$ in Figure 7.5.

A lower limit for the stellar mass range may also be inferred from the (log T_c, log ρ_c) diagram. The hydrogen burning threshold does not extend to temperatures below a few times 10^6 K (see Figure 7.2). The highest value of M for which Ψ_M still touches this threshold, before bending into the degeneracy pressure zone, may be regarded as the lower stellar mass limit, marked $\Psi_{M_{min}}$ in Figure 7.5. Objects of mass below this limit will never ignite hydrogen nor any other nuclear fuel; they will first contract and heat up and will then contract more slowly while cooling off. Such objects do not fit into our definition of stars (cf. Chapter 1). Based on Figure 7.5, the lower mass limit for a star is somewhat below $0.1M_\odot$.

When the hydrogen supply in the stellar core is finally exhausted, the star loses energy again, and the core contracts and heats up. The central point resumes its climb up the Ψ_M path. For low-mass stars, Ψ_M will soon cross the degeneracy pressure boundary and bend to the left into a horizontal line. The pressure exerted by the degenerate electron gas has become sufficient for counteracting gravity. The contraction slows down and the star cools while radiating the accumulated thermal energy, tending to a constant density (and radius), determined by M. The higher the mass, the higher will be the final density and the lower the final radius.

For higher M, Ψ_M will cross the next nuclear burning threshold. Helium now ignites in the core and another phase of thermal equilibrium is established, marking the beginning of another pause in the journey of the central point. We note that among the Ψ_M paths that cross the helium burning threshold, those corresponding to low masses do so very close to the degeneracy boundary. We may expect to encounter some form of thermonuclear instability in stars of such mass.

The story repeats itself after the exhaustion of helium: the lower-mass stars among those which have burnt helium contract, develop electron-degenerate cores, and start cooling. Contraction stops when the final density (and radius) is reached, as determined by the mass M. We have thus identified two classes of compact cooling stars: one including stars of very low mass, made predominantly of helium, and another including more massive stars composed (at least partly) of helium burning products, carbon and oxygen.

The dividing line between stars that eventually become degenerate and cool off as compact objects and those which remain in the ideal gas state due to their high temperatures, even when reaching high densities, corresponds, as we have seen, to $\Psi_{M_{Ch}}$. In principle, for $M = M_{Ch}$ contraction may go on indefinitely on the borderline of dynamical stability. However, $\Psi_{M_{\text{Ch}}}$ crosses the carbon burning threshold very near the degeneracy zone. This indicates that carbon ignites in a highly degenerate material. Nuclear burning is thermally unstable under such circumstances (see Section 6.2), and should result in a thermonuclear runaway – or *carbon detonation* – that is bound to have cataclysmic consequences. Such a fate is, however, only hypothetical for the single isolated stars of fixed mass that we are considering, since the probability of a star's having been born with a mass of (or almost) M_{Ch} is negligibly small. (In reality, stellar collapse due to carbon detonation is known to occur, as a result of mass exchange in a close binary system.)

For M above the critical value of $\sim 1.46\, M_\odot$, the central point will continue its journey up Ψ_M, stopping temporarily when a nuclear burning threshold is crossed. Note that all Ψ_M paths with $M > M_{\text{Ch}}$ terminate at the unstable iron photodisintegration boundary. Thus massive stars undergo contraction and heating phases alternating with thermal equilibrium burning of heavier and heavier nuclear fuels until their cores consist of iron. Further heating of the iron inevitably leads to its photodisintegration, which is a highly unstable process. We therefore expect the life of these stars to end in a catastrophe!

For very large M, the Ψ_M paths enter the instability zone before crossing the burning thresholds of heavy elements. Thus very massive stars are expected to be extremely short-lived, developing pair production instability that should result in a catastrophic event at early stages of their lives. In conclusion, two main types of catastrophic events are expected to terminate the lives of relatively massive stars: carbon detonation and iron photodisintegration (and, possibly, a third – caused by pair production). We may note, in passing, that the paths leading to carbon detonation and iron photodisintegration meet at the upper right corner of the $(\log T_c, \log \rho_c)$ diagram (since collapse caused by the latter implies an almost vertical ascent of Ψ_M within the instability strip). Thus the outcomes of the two different types of instability should have a great deal in common. We shall return to this speculative point later on.

To summarize, stellar masses are confined to a range spanning about three orders of magnitude, between $M_{\text{min}} \sim 0.1 M_\odot$ and $M_{\text{max}} \sim 100 M_\odot$. *All* stars undergo hydrogen burning at their centres, and since hydrogen is the most potent of the nuclear fuels, we expect central hydrogen burning to be the most common and long-lived state of stars in general. Evolution following hydrogen exhaustion proceeds differently for stars of different masses. Those under the critical mass $M_{\text{Ch}} \sim 1.46 M_\odot$ contract and cool off either after the completion of hydrogen burning or after the completion of helium burning. In stars near the critical mass, carbon detonation leads to thermonuclear instability that should end in collapse. Stars above the critical mass undergo all the nuclear burning processes, ending with iron synthesis. Subsequent heating of the iron core develops into a highly unstable state, expected to end in a catastrophic collapse or explosion (or both). Stars of very high mass may reach dynamical instability sooner, due to pair production.

How does this picture relate to the realm of observed stars? The most common among observed stars are the *main-sequence* ones, mentioned in Chapter 1. May we deduce that these are stars burning hydrogen in their cores? To prove this inference, we have to show that for hydrogen burning stars a correlation exists between luminosity and effective temperature, of the kind that defines the main sequence in the H–R diagram. This will be done in the next section. The identification of the compact cooling stars, corresponding to the horizontal part of Ψ_M tracks, with the observed *white dwarfs* is quite straightforward. It will be pursued in more detail in Chapter 8. Indeed, according to observations, one distinguishes between two types of white dwarfs: low-mass ones and more massive ones; also, the two types differ in composition, although the connection between the observed surface composition and that of the interior is debatable. Where do *red giants* fit into this picture? May we guess that they should be associated with that phase of evolution following hydrogen exhaustion, when the core contracts toward the next core-burning episode? This tricky question will be addressed shortly (Section 7.5).

Finally, observed stellar explosions – *supernovae* – are of two distinct types, termed *Type I* and *Type II*, with possible subdivisions (*Type Ia, Ib*, etc.). One may be tempted, even at this early stage in our understanding of stellar evolution, to associate one type with carbon detonation and the other with iron photodisintegration. By analyzing in more detail the properties of each, observationally as well as theoretically, we shall show in Chapter 9 that this, indeed, is the case.

7.4 The theory of the main sequence

Observationally, the main sequence is defined by an empirical relation between the luminosity and the effective temperature of a group of stars called, accordingly, main-sequence stars. This relation has the form

$$\log L = \alpha \log T_{\text{eff}} + \text{constant}, \tag{7.14}$$

where the slope α is shallower at the lower end (low L) and becomes steeper at large L. Another property of main-sequence stars is an apparent correlation between mass and luminosity, also in the form of a power law:

$$L \propto M^\nu \tag{7.15}$$

[see equation (1.6) in Chapter 1]. Our hypothesis based on theory is that main-sequence stars are those stars that burn hydrogen in their cores, their centres lying along the hydrogen-burning threshold in Figure 7.5, where the paths Ψ_M intersect this threshold. We therefore have to prove that for such stars a correlation of the type (7.14) exists and an additional one between mass and luminosity, like correlation (7.15).

7.4 The theory of the main sequence

Consider stars that have begun burning hydrogen at the centre and are in thermal and hydrostatic equilibrium. We may take their composition to be uniform throughout, equal to the initial composition that we have already assumed to be shared by all stars (see Chapter 1). Provided they are in radiative equilibrium, their structure is described by equations (5.1)–(5.7). With the further assumptions of (a) negligible radiation pressure and (b) an analytic opacity law (for the sake of simplicity, we shall adopt a constant opacity), these equations become

$$\frac{dP}{dm} = -\frac{Gm}{4\pi r^4} \tag{7.16}$$

$$\frac{dr}{dm} = \frac{1}{4\pi r^2 \rho} \tag{7.17}$$

$$\frac{dT}{dm} = -\frac{3}{4ac}\frac{\kappa}{T^3}\frac{F}{(4\pi r^2)^2} \tag{7.18}$$

$$\frac{dF}{dm} = q_0 \rho T^n \tag{7.19}$$

$$P = \frac{\mathcal{R}}{\mu}\rho T, \tag{7.20}$$

to be solved for $r(m)$, $P(m)$, $\rho(m)$, $T(m)$, and $F(m)$ in the range $0 \le m \le M$, for any value of the mass M, which is the only free parameter. Is it possible to learn something about the characteristics of these solutions without actually solving this complicated set of nonlinear differential equations? As in other cases of complex physical systems, a great deal may be learned from the *dimensional analysis* of the equations.

First, we define a dimensionless variable x, the fractional mass:

$$x = \frac{m}{M}. \tag{7.21}$$

The functions $r(m)$, $P(m)$, $\rho(m)$, $T(m)$, and $F(m)$ may then be replaced by dimensionless functions of x: $f_1(x)$, $f_2(x)$, and so forth, by the following definitions:

$$r = f_1(x)R_\star \tag{7.22}$$

$$P = f_2(x)P_\star \tag{7.23}$$

$$\rho = f_3(x)\rho_\star \tag{7.24}$$

$$T = f_4(x)T_\star \tag{7.25}$$

$$F = f_5(x)F_\star, \tag{7.26}$$

where the starred coefficients have the dimensions of the original functions, respectively.

Next, by substituting relations (7.21)–(7.23) into equation (7.16), we obtain

$$\frac{P_\star}{M}\frac{df_2}{dx} = -\frac{GMx}{4\pi f_1^4 R_\star^4}. \qquad (7.27)$$

In a physical equation the dimensions on the two sides must match, and hence in equation (7.27), where x, f_1, and f_2 are dimensionless, $P_\star$ must be proportional to $GM^2/R_\star^4$. Adopting (without loss of generality) a proportionality constant of unity, equation (7.27) may be separated into

$$\frac{df_2}{dx} = -\frac{x}{4\pi f_1^4} \qquad P_\star = \frac{GM^2}{R_\star^4}, \qquad (7.28)$$

and repeating the procedure for equation (7.17), then equations (7.20), (7.18), and (7.19),

$$\frac{df_1}{dx} = \frac{1}{4\pi f_1^2 f_3} \qquad \rho_\star = \frac{M}{R_\star^3} \qquad (7.29)$$

$$f_2 = f_3 f_4 \qquad T_\star = \frac{\mu P_\star}{\mathcal{R} \rho_\star} \qquad (7.30)$$

$$\frac{df_4}{dx} = -\frac{3 f_5}{4 f_4^3 (4\pi f_1^2)^2} \qquad F_\star = \frac{ac}{\kappa}\frac{T_\star^4 R_\star^4}{M} \qquad (7.31)$$

$$\frac{df_5}{dx} = f_3 f_4^n \qquad F_\star = q_0 \rho_\star T_\star^n M. \qquad (7.32)$$

On the left of equations (7.28)–(7.32) we have a set of nonlinear differential equations, now independent of M, for the variable functions f_{1-5} that have been defined in the range $0 \leq x \leq 1$ by equations (7.22)–(7.26). The dimensional coefficients that appear on the right-hand side of equations (7.22)–(7.26) are obtained as *functions of the stellar mass M* by solving the set of *algebraic* equations on the right of equations (7.28)–(7.32). Combining the solutions of the differential equations and the algebraic equations, we may obtain from equations (7.22)–(7.26) the profiles of any physical characteristic (temperature, density, pressure, etc.) for any value of M. The important conclusion is that the shape of the profiles as a function of the fractional mass is the same in all stars, the profiles differing only by a constant factor determined by the mass. This similarity property is called *homology*.

By solving only the simple set of relations between the starred quantities, we may therefore derive the dependence of physical properties on the stellar mass, without actually solving the differential equations. Substituting equations (7.28)

7.4 The theory of the main sequence

and (7.29) into equation (7.30), we obtain

$$T_\star = \frac{\mu G}{\mathcal{R}} \frac{M}{R_\star}, \qquad (7.33)$$

and inserting this relation, in turn, into equation (7.31), we have

$$F_\star = \frac{ac}{\kappa} \left(\frac{\mu G}{\mathcal{R}}\right)^4 M^3. \qquad (7.34)$$

Thus fluxes at a given fractional mass in stars of different masses relate as the cube of the mass ratio. For example, the radiative flux across a spherical surface enclosing, say, half the total mass will be a thousand times larger in a star of $10 M_\odot$ than in a star of $1 M_\odot$. The same applies to any other value of x. In particular, the surface ($x = 1$) flux, or the luminosity, will be proportional to the mass cubed,

$$L \propto M^3. \qquad (7.35)$$

This is the desired relation between luminosity and mass, to be compared to that derived observationally for main-sequence stars (see below). We recall that a similar relation emerged from the simple *standard model* discussed in Chapter 5. Combining equations (7.34) and (7.32) and substituting equations (7.28)–(7.30) yields the dependence of $R_\star$ on the mass M in the form

$$R_\star \propto M^{\frac{n-1}{n+3}}, \qquad (7.36)$$

which relates radii corresponding to a given fractional mass in stars of different masses. This holds, in particular, for $x = 1$, that is, for the stellar radius R. Hence, for a large n, such as $n \approx 16$ corresponding to CNO-cycle hydrogen burning, the radius will be almost proportional to the mass. For $n = 4$ that approximates hydrogen burning by the p–p chain, the dependence is weaker, $R \propto M^{3/7}$. We note that in all cases the radius increases with increasing mass, in contrast to compact degenerate stars (white dwarfs), where the radius is inversely proportional to some power of the mass. The power in relation (7.36) is always smaller than unity (tending to 1, in principle, as $n \to \infty$). Inserting relation (7.36) into equation (7.29), we obtain the variation of density with mass M:

$$\rho_\star \propto M^{2\frac{3-n}{3+n}}. \qquad (7.37)$$

Since $n > 3$, the density decreases with increasing stellar mass. Thus stars of low mass are denser than massive stars at any x, again in contrast to degenerate stars. That this holds for the stellar centre ($x = 0$) is obvious from Figure 7.4.

Exercise 7.1: Derive the dependence of the pressure $P_\star$ and of the temperature $T_\star$ on M (a) in general form; (b) for $n = 4$ and $n = 16$.

We are now ready for the crucial test of our hypothesis that equations (7.16)–(7.20) may be taken to describe main-sequence stars. In the relation between luminosity and effective temperature $L = 4\pi R^2 \sigma T_{\text{eff}}^4$ [equation (1.3)] the radius R may be eliminated, using relations (7.35) and (7.36), to obtain

$$L^{1-\frac{2(n-1)}{3(n+3)}} \propto T_{\text{eff}}^4. \qquad (7.38)$$

Taking logarithms on both sides, we have for $n = 4$

$$\log L = 5.6 \log T_{\text{eff}} + \text{constant}, \qquad (7.39a)$$

while for $n = 16$

$$\log L = 8.4 \log T_{\text{eff}} + \text{constant}. \qquad (7.39b)$$

These are the calculated slopes for the lower part (low L and M) and for the upper part (high L and M) of the main sequence in the ($\log L$, $\log T_{\text{eff}}$) diagram, in agreement with those derived observationally.

Other characteristics of the main sequence are also readily explained. The nuclear energy reservoir of a star is, obviously, proportional to its mass. In thermal equilibrium the rate of consumption of the nuclear fuel is equal to the rate of energy release L. Hence the duration of the main-sequence (hydrogen burning) stage, τ_{MS}, should roughly satisfy

$$\tau_{\text{MS}} \propto \frac{M}{L} \propto M^{-2}, \qquad (7.40)$$

where we have used relation (7.35). The larger the stellar mass, the shorter the time spent by the star on the main sequence (burning hydrogen). This explains why in an ensemble of stars born at the same time, the more massive among them leave the main sequence earlier. With the passage of time, the main sequence of this ensemble becomes gradually shorter, as stars that are less and less massive leave it. This is the reason for the different extent (or upper end) of main sequences corresponding to stellar clusters of different ages, as we have encountered in Chapter 1.

We have concluded on the basis of the schematic picture of stellar evolution (Figure 7.5) that there should be a minimal mass for stars capable of igniting hydrogen. We may now attempt to calculate it more accurately. According to equation (7.33) and relation (7.36), the temperature within stars of different masses varies as $M/R_\star \propto M^{4/(n+3)}$, that is, as M to a positive power. This holds, in

7.4 The theory of the main sequence

particular, for the central temperature (the highest temperature within a main-sequence star), which thus decreases with decreasing M,

$$T_c \propto M^{4/7}, \tag{7.41a}$$

where we have substituted $n = 4$, appropriate to low stellar masses (low temperatures). The lowest temperature required for hydrogen burning into helium is $T_{\min} \approx 4 \times 10^6$ K, applying to the p–p chain. We know that the Sun is a main-sequence star burning hydrogen predominantly via the p–p chain, from its location in the H–R diagram and from detailed studies of its interior. We may therefore calibrate relation (7.41a):

$$\frac{T_c}{T_{c,\odot}} = \left(\frac{M}{M_\odot}\right)^{4/7}. \tag{7.41b}$$

The condition for hydrogen ignition

$$T_c \geq T_{\min}$$

may thus be translated into a condition on the mass

$$\frac{M}{M_\odot} \geq \left(\frac{T_{\min}}{T_{c,\odot}}\right)^{7/4}, \tag{7.42}$$

yielding $M_{\min} \approx 0.1 M_\odot$ for the estimated $T_{c,\odot} \approx 1.5 \times 10^7$ K. The luminosity corresponding to this mass may be calculated from relation (7.35) after calibrating it with the aid of $L_\odot$ and $M_\odot$:

$$\frac{L_{\min}}{L_\odot} = \left(\frac{M_{\min}}{M_\odot}\right)^3 \approx 10^{-3}, \tag{7.43}$$

thus defining the lower end of the main sequence.

Exercise 7.2: Calculate the effective temperature corresponding to the lower end of the main sequence.

Exercise 7.3: Using the condition $L \leq L_{\text{Edd}}$ [with L_{Edd} given by equation (5.37)], derive an upper limit for the mass and the luminosity of main-sequence stars. Estimate the effective temperature at the upper end of the main sequence.

> **Exercise 7.4:** Find the relation between L and M and the slope of the main sequence, assuming an opacity law $\kappa = \kappa_0 \rho T^{-7/2}$ (the Kramers opacity law) and $n = 4$.

We now return briefly to the mass-luminosity relation (7.35). Generally, the power depends on the adopted opacity law as well as on n, although for the constant opacity we have assumed – appropriate for electron scattering, which dominates at high temperatures – the power of 3 is independent of n. For the Kramers opacity law, appropriate for relatively low temperatures, the power is $\frac{5+31/2n}{1+5/2n}$ (Exercise 7.4), which is close to 5 for $n \geq 4$. As we have seen that the temperature scales with the mass, this explains the changing slope of the observed mass-luminosity relation (Figure 1.6) from 5 on the lower part to 3 on the upper part. The slopes of the main sequence, as derived earlier, would also change to some extent for a different opacity law, but the upper part would still remain much steeper than the lower one.

> **Exercise 7.5:** Repeat the dimensional analysis and derive its results for fully convective stars, where structure equations (7.18) and (7.20) are replaced by
>
> $$P = K_a \rho^{\gamma_a}$$
> $$T = K'_a P^{(\gamma_a - 1)/\gamma_a}$$
>
> (see Sections 6.5–6.6), assuming $\gamma_a = 5/3$ and the same K_a (and K'_a) for these stars.

In conclusion, we have succeeded to explain most features of the observed main sequence. Furthermore, if we take into account that the initial composition is not strictly the same for all stars, we also understand why the main sequence is a strip rather than a line. The hypothesis that main-sequence stars are those stars that burn hydrogen in a relatively small core has thus turned into *the theory of the main sequence*. Strictly, the theory applies to the *zero-age main sequence*, when the composition is truly homogeneous. And we should bear in mind that it ignores convection. But why can the simple and quite general procedure employed be applied only to hydrogen burning? The reason why it cannot be applied to other types of stars (other stages of evolution) is that the crucial homogeneity assumption – both of composition and of physical state – is no longer valid at advanced evolutionary stages.

7.5 Outline of the structure of stars in late evolutionary stages

The same basic diagram that was used to describe the *evolution* of stars may also serve to describe the *structure* of a star at a given evolutionary stage. Consider a

7.5 Outline of the structure of stars in late evolutionary stages

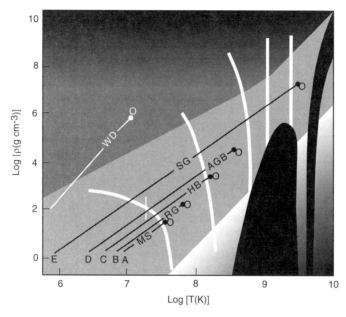

Figure 7.6 Schematic illustration of the stellar configuration in different evolutionary phases for a $10M_\odot$ star (A, B, C, D, E) and a white dwarf (WD).

star of mass M: for any given point m within it we have the value of the local temperature $T(m)$ and the value of the local density $\rho(m)$, which define a point in the $(\log T, \log \rho)$ plane. Joining the points corresponding to different values of m between 0 and M, we obtain a parametric line that traces the structure of the star in the $(\log T, \log \rho)$ plane. One end of the line – the central point – lies on the Ψ_M curve; the other – the surface – is characterized by a temperature considerably below 10^6 K and by a very low density. Hence structure lines run across the $(\log T, \log \rho)$ plane toward the lower left corner. The exact shape of these lines may be complicated (only polytropes would be described by straight lines on logarithmic scales) and may change with time, as the central point moves along Ψ_M. Nevertheless, they will invariably lead (more or less) monotonically from the central point toward low temperatures and densities.

An example is given, schematically, in Figure 7.6 for a star of $10M_\odot$ by a series of structure lines labelled A, B, C, ... with origins at a chronological series of points (labelled O) lying along the evolutionary track Ψ_{10} of Figure 7.5. These lines may be taken to roughly represent the evolving structure of the star. We recognize line A as outlining the main-sequence structure (MS). The next line, B, describes the star at a later stage when hydrogen has been depleted in the core. We note that the conditions for hydrogen burning are now fulfilled at some point outside the core, where A intersects the hydrogen burning threshold. Thus hydrogen burning continues in a *shell* outside the helium core. The relatively cool region beyond the burning shell constitutes a chemically homogeneous *envelope*. The core itself, now devoid of energy sources, is contracting and heating up.

Regarding this particular evolutionary phase, Martin Schwarzschild wrote in his book:

> "It would thus clearly be safer if we stopped our discussion of stellar evolution here and waited for the results from the big computers, which we may expect in the nearest future. But for those whose curiosity is stronger than their wish for safety we shall go on – fully aware of the risk."
>
> Martin Schwarzschild: Structure and Evolution of the Stars, 1958

So we, too, shall take the risk and go on, the results of numerical calculations awaiting us in the next chapter.

Assuming the contraction of the core to occur quasi-statically, on a timescale which is much longer than the dynamical timescale, the virial theorem may be assumed to hold. Provided the amount of energy gained (or lost) during this phase is negligible with respect to the total stellar energy (that is, thermal equilibrium is maintained), the latter may be assumed to remain constant. As discussed in Section 2.8, under such circumstances the gravitational potential energy and the thermal energy are each conserved. Consequently, contraction of the core must be accompanied by expansion of the envelope, so as to conserve the gravitational potential energy. At the same time, heating of the core must result in cooling of the envelope, for the thermal energy to be conserved. In particular, the surface (effective) temperature drops, the blackbody radiation thus shifting to the red. The star assumes the appearance of a *red giant* (RG).

In order to get a rough idea of the amount of expansion that might take place, we may do a very simple exercise: consider two equal mass elements Δm_1 and Δm_2 at a distance r_0 from the centre of a star and regard $m(r_0)$ as a point mass. Suppose that one element moves toward the centre, to a distance r_1, and the other outward, to a distance r_2, so that the gravitational energy of the system is conserved. It is easily verified that the distances measured in units of r_0, $\tilde{r}_1 = r_1/r_0$, and $\tilde{r}_2 = r_2/r_0$ are related by $\tilde{r}_2 = (2 - \tilde{r}_1^{-1})^{-1}$. We find that when one element moves inward $\sim$10% of r_0 ($\tilde{r}_1 = 0.9$), the other moves outward by about the same amount ($\tilde{r}_2 = 1.13$). When the inward displacement is 20% ($\tilde{r}_1 = 0.8$), however, the outward one is more than 30% ($\tilde{r}_2 = 1.33$), and the difference increases, r_2 tending to *infinity*, as r_1 approaches *half* the original distance. This exercise should not be taken too literally: the gravitational energy is conserved globally, not by separate mass elements; the motion occurs within the mass of the star and not outside it, and so forth. Nevertheless, the general conclusion that it was meant to emphasize – that a moderate amount of core contraction may entail a significant expansion of the envelope – is true.

If the total energy does not remain constant as assumed, but rather increases ($L_{\rm nuc} > L$ on the average), then it is easy to see that the effect of envelope expansion upon core contraction will be all the more considerable. Therefore the giant dimensions that red giants may reach should not surprise us. We note,

7.5 Outline of the structure of stars in late evolutionary stages

however, that if the total energy of the star decreases while the core contracts ($L_{\text{nuc}} < L$), we cannot draw any definite conclusion: the envelope may then either expand, or remain unchanged, or even contract too, depending on the difference between the energy drop resulting from core contraction and the overall energy drop ($L_{\text{nuc}} - L$). Only detailed stellar evolution computations can provide the answer as to the departure from thermal equilibrium (its trend and extent). But provided $L_{\text{nuc}} \geq L$, we may safely claim that a star with a contracting core should evolve into a red giant.

> **Note:** The first detailed calculations of evolving inhomogeneous stellar models, carried out by Allan Sandage and Martin Schwarzschild in 1952, indeed showed this effect. The models consisted of a contracting core and an envelope, separated by a hydrogen burning shell. It was found that "*...as the cores contract, the envelopes greatly expand. Thus from the initial configuration, which is near the main sequence, the stars evolve rapidly to the right in the H–R diagram...*"
>
> It is interesting to note that these calculations were done during the brief period of time between Salpeter's solution for the 3α process and Hoyle's prediction of its resonant character (see Section 4.5). Thus, at the time, the estimated threshold temperature for helium ignition was $\sim 2 \times 10^8$ K. Sandage and Schwarzschild found, to their disappointment, that while the cores contract and heat up toward helium ignition, the envelopes expand way beyond the observed red giant branch. They concluded, or rather speculated, that when the central temperature reaches 1.1×10^8 K, "*...a physical process not included in the present computations should start to play an essential role...,*" so as to halt the contraction of cores and expansion of envelopes. This could have been a second, independent argument for postulating a resonant energy level in the carbon nucleus. Indeed, this level reduces the threshold temperature for the 3α process to about 10^8 K!

When the helium ignition temperature is finally reached at the centre, core contraction stops. The structure of the star is described by line C in Figure 7.6. Two energy sources are now exploited: the main one, helium burning in the core, and a secondary one, hydrogen burning in a shell around it. When helium is exhausted in the core, another phase of core contraction and envelope expansion sets in. Since the core is now more condensed, envelope expansion is even more pronounced, turning the star into a *supergiant*. The structure of the star at this point is described by line D in Figure 7.6: outside the carbon-oxygen core resulting from helium burning, we find *two* burning shells, where D intersects the helium burning threshold and the hydrogen burning threshold, respectively. The composition of the star is stratified: enveloping the carbon-oxygen core is a helium layer, with the helium burning shell between them. The outer boundary of the helium layer is defined by the hydrogen burning shell, which separates the helium layer from the hydrogen-rich envelope. The hydrogen burning shell feeds fresh fuel to the helium burning one, and so both advance outward. The process is quite complicated in detail, as we shall see in the next chapter (where symbols HB – horizontal branch – and AGB will be explained). Finally, when all the nuclear processes are over in the stellar core, the structure of the star, line E in Figure 7.6, is layered like an onion's, each layer having a different composition, with lighter elements lying above heavier ones. The supergiant (SG) is now a *supernova progenitor*. The spectacular, albeit brief remainder of its evolutionary course will be discussed in Chapter 9.

A similar chain of arguments may be applied to stars of other masses.

7.6 Shortcomings of the simple stellar evolution picture

It is noteworthy that the very first rough sketch of the global evolution of stars was outlined by Bethe in 1939(!); this is how Bethe ended his treatise on energy production in stars, which paved the way to the modern theory of stellar evolution:

> "... It is very interesting to ask what will happen to a star when its hydrogen is almost exhausted. Then, obviously, the energy production can no longer keep pace with the requirements of equilibrium so that the star will begin to contract. Gravitational attraction will then supply a large part of the energy. The contraction will continue until a new equilibrium is reached. For 'light' stars of mass less than $6\mu^{-2}$ sun masses, the electron gas in the star will become degenerate and a white dwarf will result. In the white dwarf state, the necessary energy production is extremely small so that such a star will have an almost unlimited life. ...
>
> For heavy stars, it seems that the contraction can only stop when a neutron core is formed. The difficulties encountered with such a core may not be insuperable in our case because most of the hydrogen has already been transformed into heavier and more stable elements so that the energy evolution at the surface of the core will be by gravitation rather than by nuclear reactions. However, these questions obviously require much further investigation."
>
> Hans A. Bethe: Physical Review, 1939

In the present chapter, we have built a frame for the theme of stellar evolution and we have outlined a more elaborate sketch (along the same basic lines!), but the picture is still far from being complete. In order to fill in the details, we shall have to rely on numerical computations of stellar evolution – the computational laboratories of stars. This will be the subject of the next chapter, but in order to assess the authenticity of our sketch, the results of complex numerical calculations for the evolution of stars of various masses, as they appear in the ($\log T_c$, $\log \rho_c$) plane, are shown in Figure 7.7. The general trend is remarkably similar to that of Figure 7.5 obtained on the basis of simple arguments. Leaving aside the deviations associated with the ignition of a nuclear fuel (in particular the expected explosive helium ignition at the centre of the $1 M_\odot$ star), we may be surprised to discover

7.6 Shortcomings of the simple stellar evolution picture

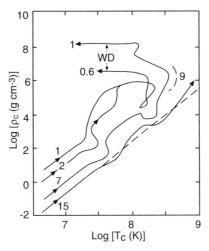

Figure 7.7 Relation of central density to central temperature as obtained from complex numerical calculations of the evolution of stars of various masses [adapted from I. Iben (1974), *Ann. Rev. Astron. Astrophys.*, 12].

that stars as massive as $7M_\odot$, and perhaps up to $9M_\odot$, end their lives as white dwarfs. We have expected this to happen only below M_{Ch}!

This points out the fallacy of our assumption concerning the conservation of the stellar mass during evolution. We should have suspected this assumption to be wrong, especially for massive stars, from the conclusions of the *standard model* (Section 5.6). Observational considerations, too, suggest that mass loss must occur. Several low-mass white dwarfs in the solar neighbourhood, with accurately determined masses ($0.4M_\odot$ or less), are long known. If stars conserved their masses, it would follow that the Galaxy is old enough for stars of $0.4M_\odot$ or less to have evolved off the main sequence. We should then expect to encounter at least some star clusters with main sequences ending at luminosities below that corresponding to a mass of $0.4M_\odot$, but no such clusters are known. In fact, the main sequences of all known clusters extend to considerably higher luminosities (corresponding to masses above $0.7M_\odot$), indicating a younger age. It would be very difficult to explain why *all* star clusters should be much younger than the Galaxy within which they reside. It is far more natural to assume that the Galaxy is about as old as its oldest clusters, which forces us to conclude that stars lose mass, particularly after leaving the main sequence.

"...We are forced to accept the short time scales for most clusters and look for processes by which a massive evolved star is able to lose a large fraction of its mass, so it can settle down into a cooling white dwarf; thus, we link the problem of the origin of white dwarfs with that of the ultimate fate of stars well above the Chandrasekhar limit."

Leon Mestel: The Theory of White Dwarfs, 1965

Nowadays, when modern telescopes are able to detect white dwarfs in dense globular clusters, this argument is even stronger: the white dwarfs have *lower* masses than main-sequence stars of *the same* cluster.

As it turns out, stars lose a significant fraction of their masses by a *stellar wind*, such as that emanating from the Sun, only much more substantially in the case of massive stars, where radiation pressure is considerable. Hence the evolutionary paths Ψ_M in Figure 7.5, should have increasingly steeper slopes, as the initial mass M increases. This means that stars initially more massive than M_{Ch} may become white dwarfs, their Ψ paths shifting quickly toward paths corresponding to lower and lower masses, the evolutionary course being very similar to that described earlier for a mass of about $1 M_\odot$. Therefore the general picture *remains valid*, except that the dividing mass between stars that will end up as white dwarfs and stars that will become supernovae is, in reality, higher than $1.46 M_\odot$. To determine how high, a model of mass loss is required; unfortunately, no simple theoretical model has been found for this phenomenon.

Exercise 7.6: Consider the hypothetical evolution of a star of initial mass M_0. The star's core grows in mass as a result of nuclear burning. The nuclear processes release an amount of energy Q per gram of burnt material. The star loses mass (by means of a stellar wind) at a rate proportional to its constant luminosity L: $\dot{M} = -\alpha L$. (a) Find the mass of the core as a function of time, $M_c(t)$, assuming that $M_c(0) = 0$. (b) Find the mass of the envelope as a function of time, $M_e(t)$, noting that $M_e(0) = M_0$. (c) What is the core mass when the envelope mass vanishes? (d) Calculate the upper limit of M_0, for which the star will become a white dwarf, given $Q = 5 \times 10^{14}$ J kg^{-1} (from turning solar composition into carbon and oxygen) and $\alpha = 10^{-14}$ kg J^{-1}.

Another process that has been neglected is neutrino emission in *dense* cores, which has a marked cooling effect. As the rise in temperature between late burning stages is impeded by neutrino cooling, the slopes of the Ψ_M curves should become somewhat steeper than 3. However, this effect does not alter any of our conclusions.

The main shortcomings of the simple picture are (a) the total lack of time spans for the different processes and (b) the ignorance of the outward appearance of the star at each stage. Both factors render a comparison with observations impossible (statistically as well as individually). Since our main purpose is to reproduce as accurately as possible the observed stellar characteristics – not only their trends – we must resort to detailed stellar models. Having acquired a basic understanding of the principles involved, we may expect a smooth sail through the ocean of evolutionary computations.

8

The evolution of stars – a detailed picture

This chapter differs from previous ones by being descriptive rather than analytical. An account will be given of the evolution of stars as it emerges from full-scale numerical calculations – solutions of the set of equations (2.54), with accurate equations of state, opacity coefficients, and nuclear reaction rates. Such numerical studies of stellar evolution date back to the early 1960s, when the first computer codes for this task were developed. The first to program the evolution of stellar models on an electronic computer were Haselgrove and Hoyle in 1956. They adopted a method of direct numerical integration of the equations and fitting to outer boundary conditions. A much better suited numerical procedure for the two-boundary value nature of the stellar structure equations (essentially a *relaxation method*) was soon proposed by Louis Henyey; it is often referred to as *the Henyey method* and it has been adopted by most stellar evolution codes to this day. Among the numerous calculations performed by many astrophysicists all over the world since the early 1960s, the lion's share belongs to Icko Iben Jr. The detailed results of such computations cannot always be anticipated on the basis of fundamental principles, and simple, intuitive explanations cannot always be offered. We must accept the fact that, being highly nonlinear, the evolution equations may be expected to have quite complicated solutions.

As the complete solutions of the evolution equations provide, in particular, the observable surface properties of stars, we shall focus in this chapter, more than we have previously done, on the comparison of theoretical results with observations. The ultimate test to the stellar evolution theory is the understanding of the H–R diagram in all its aspects (described briefly in Chapter 1). We thus expect to find stars in the H–R diagram where theoretical models predict them to be. Moreover, the basic statistical principle mentioned in Chapter 1 should apply: the longer an evolutionary phase of an individual star, the larger the number of stars to be observed in that particular phase. A detailed comparison between theoretical predictions and observations is thus possible for long evolutionary phases, such as core hydrogen burning and, to a lesser extent, core helium burning. Proceeding to advanced evolutionary stages, neutrino emission from the dense, hot cores of massive stars, acting as an efficient energy removing agent, accelerates the rate of evolution by requiring an enhanced rate of nuclear energy supply. Hence the weak nuclear fuel (from carbon to iron, the amount of energy release per unit mass of burnt material is relatively small) is quickly consumed. Consequently, the probability of detecting stars during these brief evolutionary phases is low. Cooling, following the completion of nuclear burning in relatively low-mass stars, is again a slow process, but cooling stars – white dwarfs – become gradually fainter and more difficult to detect.

8.1 The Hayashi zone and the pre-main-sequence phase

The previous chapter dealt with the evolution of stars by following the path of the stellar centre in the $(\log T, \log \rho)$ plane. The present chapter, focusing on the stellar surface, follows evolutionary tracks in the $(\log L, \log T_{\text{eff}})$ plane, the theorists' H–R diagram. In the $(\log T, \log \rho)$ plane we found zones of instability, which have constrained the evolutionary paths of stars. We shall now show that the $(\log L, \log T_{\text{eff}})$ plane has its own "forbidden zone." It is known as the *Hayashi forbidden zone* and its boundary as the *Hayashi track*, after Chushiro Hayashi, who was the first to point out and study this type of instability in the early 1960s. The forbidden zone's boundary is determined by the hypothetical evolution of a fully convective star.

Consider a fully convective star of mass M, where convection reaches out to the stellar photosphere. In Section 6.6 we showed that in a convective zone the temperature gradient is very closely adiabatic. On the one hand, even a slight superadiabaticity gives rise to high heat fluxes which reduce the temperature gradient. On the other hand, subadiabaticity quenches convection and reduces the heat flux; as a result, the temperature gradient steepens. Therefore, if convection persists, the temperature gradient remains very close to the adiabatic. Neglecting the mass and thickness of the photosphere with respect to the stellar mass M and radius R, we may adopt a very simple description for the interior of a fully convective star as a polytrope of index $n = (\gamma_a - 1)^{-1}$,

$$P = K\rho^{1+\frac{1}{n}} \tag{8.1}$$

(see Section 5.3). The coefficient K is related to M and R by equation (5.23):

$$K^n = C_n G^n M^{n-1} R^{3-n}, \tag{8.2}$$

the constant C_n depending on n, $C_n = \frac{4\pi}{(n+1)^n} \frac{R_n^{n-3}}{M_n^{n-1}}$. We have one free parameter, the value of R, which will be determined by joining the fully convective interior to the radiative photosphere at the boundary $r = R$. The ability of the

8.1 The Hayashi zone and the pre-main-sequence phase

photosphere to radiate the energy flux crossing this boundary will depend on the change in density, temperature, and pressure across it. Hydrostatic equilibrium requires

$$\frac{dP}{dr} \approx -\rho \frac{GM}{R^2}, \tag{8.3}$$

and integrating from R, where the pressure is P_R, to the point where the pressure vanishes, or, for simplicity, to infinity, we obtain

$$P_R = \frac{GM}{R^2} \int_R^\infty \rho \, dr. \tag{8.4}$$

The temperature at R is the effective temperature of the star, satisfying $L = 4\pi R^2 \sigma T_{\text{eff}}^4$. The optical depth of the photosphere is of the order of unity (see Section 3.7), the exact value depending on the type of solution of the radiative transfer equation. Thus $\int_R^\infty \kappa \rho \, dr = \bar{\kappa} \int_R^\infty \rho \, dr = 1$, where $\bar{\kappa}$ is the opacity averaged over the photosphere. Taking $\bar{\kappa}$ to be the opacity at R and expressing it as a power law in density ρ_R and temperature T_{eff}, as in equation (3.63), we have, as a crude approximation,

$$\kappa_0 \rho_R^a T_{\text{eff}}^b \int_R^\infty \rho \, dr = 1. \tag{8.5}$$

Combining equations (8.4) and (8.5) we obtain

$$P_R = \frac{GM}{R^2 \kappa_0} \rho_R^{-a} T_{\text{eff}}^{-b}. \tag{8.6}$$

A further relation among pressure, density, and temperature at R is provided by the equation of state, for which we adopt the simplest case of an ideal gas and negligible radiation pressure, equation (3.28), $P_R = \frac{\mathcal{R}}{\mu} \rho_R T_{\text{eff}}$. We thus arrive at a set of four equations, all in the form of products of powers of physical quantities, which are easily solved when turned into linear logarithmic equations:

$$\log P_R = \log M - 2 \log R - a \log \rho_R - b \log T_{\text{eff}} + \text{constant} \tag{8.7}$$

$$n \log P_R = (n-1) \log M + (3-n) \log R + (n+1) \log \rho_R + \text{constant} \tag{8.8}$$

$$\log P_R = \log \rho_R + \log T_{\text{eff}} + \text{constant} \tag{8.9}$$

$$\log L = 2 \log R + 4 \log T_{\text{eff}} + \text{constant}. \tag{8.10}$$

By eliminating $\log R$, $\log \rho_R$, and $\log P_R$, we obtain a relation among $\log L$,

$\log T_{\rm eff}$, and $\log M$, in the form

$$\log L = A \log T_{\rm eff} + B \log M + {\rm constant}, \tag{8.11}$$

$$A = \frac{(7-n)(a+1) - 4 - a + b}{0.5(3-n)(a+1) - 1}, \qquad B = -\frac{(n-1)(a+1) + 1}{0.5(3-n)(a+1) - 1}, \tag{8.12}$$

which traces a line in the ($\log L$, $\log T_{\rm eff}$) diagram, the Hayashi track, for each value of M. These tracks play a similar role to that of the Ψ_M tracks in the ($\log T$, $\log \rho$) plane, but they cannot be taken to represent evolutionary paths, as their Ψ_M counterparts, because the assumptions on which they were derived are not generally valid. They represent *asymptotes* to evolutionary paths, as we shall show shortly.

To simplify the discussion, we assume $a = 1$, which is a reasonable approximation. The power b, however, may assume a wide range of values, mostly positive (as seen from Figure 3.3), because photospheric temperatures are relatively low. The coefficients (8.12) thus reduce to

$$A = \frac{9 - 2n + b}{2 - n}, \qquad B = -\frac{2n - 1}{2 - n}. \tag{8.13}$$

In addition, we recall that the polytropic index is constrained by dynamical stability to $n < 3$ (Section 6.3), and hence overall $1.5 \le n < 3$. The first conclusion to be drawn from equations (8.11) and (8.13) is that the slope of the Hayashi track is extremely steep. For $b = 4$ and $n = 1.5$, typical of low temperatures, we obtain $A = 20$ – an almost vertical line. Consequently, tracks corresponding to different stellar masses are close to each other, with larger values of M shifting the lines always to the left. This is because ${\rm sgn}[B] = -{\rm sgn}[A]$, and hence for $A > 0$ a higher M lowers the line, while for $A < 0$, a higher M raises it. (We recall that in the H–R diagram, the effective temperature increases leftward.) We further note that the slope changes sign for $n > 2$. Thus, for example, the slope will be differently inclined for a photosphere of atomic hydrogen ($n = 1.5$) and one of molecular hydrogen ($n = 2.5$). Accurate calculations of convective stellar models show that the Hayashi track corresponding to a given M changes slope, bending slightly to the left at low L and shifting to the right at large L.

In order to understand the significance of Hayashi tracks, we characterize a star by a unique value $\bar{\gamma}$, obtained by averaging $\gamma = \frac{d \ln P}{d \ln \rho}$ over the entire star. Similarly, $\bar{\gamma}_a$ denotes the average adiabatic exponent. For a fully convective star, we obviously have $\bar{\gamma} = \bar{\gamma}_a$. If the star has radiative zones, then in some regions $\gamma < \gamma_a$, and hence $\bar{\gamma} < \bar{\gamma}_a$. The corresponding average polytropic index n for such stars satisfies $n > n_a$, where n_a is taken to denote the adiabatic polytropic index that defines the Hayashi track. Therefore, $\bar{\gamma} > \bar{\gamma}_a$, or $n < n_a$, can only arise from superadiabaticity, which is unstable and hence "forbidden." Now, consider a star of mass M and

8.1 The Hayashi zone and the pre-main-sequence phase

luminosity L, whose configuration can be described, as above, by a polytrope overlaid by a photosphere. How would its effective temperature change with the polytropic index n? To answer this question, we have to reconstruct equations (8.7)–(8.10), taking account of the dependence of the constants (on the right-hand sides) on the polytropic index in order to obtain the function $\log T_{\mathrm{eff}}(n)$. When this tedious task is accomplished, the result is $d \log T_{\mathrm{eff}}/dn > 0$. Consequently, the forbidden zone corresponding to $n < n_a$ lies *to the right* of the Hayashi tracks in the H–R diagram.

The role of Hayashi tracks and forbidden zone is best illustrated by the pre-main-sequence evolution of stars. The very beginning of a star's life is marked by a rapid collapse of an unstable gaseous cloud. The initiation of such a collapse being a galactic, rather than a stellar process, will be discussed in Chapter 10. At first, the material is transparent, but as it condenses and its temperature rises, it eventually becomes opaque. The interior is now shielded and the boundary layer from which radiation escapes defines a discernible object which will become a star. This occurs at densities of about 10^{-10}–10^{-9} kg m^{-3} and temperatures of a few hundred degrees Kelvin. Under such conditions hydrogen is in molecular (H$_2$) form. The gas is too cold to resist the gravitational force, and contraction proceeds, essentially as radial free fall, on the dynamical timescale [of the order of $1/\sqrt{G\rho}$; equation (2.57)]. We note that this timescale is considerably longer than that typical of mature stars, densities being so much lower. The rising gas temperature becomes, eventually, high enough for dissociation of the hydrogen molecules to take place, then for ionization of the hydrogen atoms and, finally, for ionization of the helium atoms. All these processes absorb a vast amount of energy, which is supplied by the gravitational energy released in contraction. The gas temperature is now prevented from further increase, much in the same way as the temperature of boiling water remains constant, although energy is continually supplied to it to keep it boiling. Thus free fall continues throughout these stages. When ionization of hydrogen and helium is almost complete, the gas temperature increases again due to the release of gravitational energy. There comes a time when it generates sufficient pressure to oppose the gravitational pull and a state of hydrostatic equilibrium is established. The gaseous condensation has now become a protostar.

A rough estimate of protostellar characteristics may be obtained by assuming that all the gravitational energy released in collapse to the protostellar radius R_{ps} practically from infinity, $\alpha GM^2/R_{\mathrm{ps}}$, was absorbed in dissociation of molecular hydrogen and ionization of hydrogen and helium, although in reality a fraction was emitted as radiation. Denoting by $\chi_{\mathrm{H_2}}$ the dissociation potential of H$_2$ (4.5 eV), by χ_{H} the ionization potential of hydrogen (13.6 eV), and by χ_{He} the total ionization potential of helium (79 eV = 24.6 eV + 54.4 eV), we have

$$\alpha \frac{GM^2}{R_{\mathrm{ps}}} \approx \frac{M}{m_{\mathrm{H}}} \left(\frac{X}{2} \chi_{\mathrm{H_2}} + X \chi_{\mathrm{H}} + \frac{Y}{4} \chi_{\mathrm{He}} \right), \tag{8.14}$$

and taking $Y \approx 1 - X$ and $\alpha \approx \frac{1}{2}$, we obtain

$$\frac{R_{ps}}{R_\odot} \approx \frac{50}{1 - 0.2X} \frac{M}{M_\odot}. \tag{8.15}$$

The protostar being in hydrostatic equilibrium, an average internal temperature may be estimated from the virial theorem, as in Section 2.4 [equation (2.29)], using approximation (8.15):

$$\bar{T} = \frac{\alpha}{3} \frac{\mu}{k} \frac{GMm_H}{R_{ps}} \approx 6 \times 10^4 \text{ K} \tag{8.16}$$

(for $X \approx 0.7$), independent of the stellar mass. At this temperature the opacity is still very high (see Figure 3.3), the flow of radiation is hindered, and hence the protostar is fully convective. This is the starting point of the Hayashi evolutionary phase. In the ($\log L$, $\log T_{\text{eff}}$) diagram the star descends along its Hayashi track at almost constant effective temperature, its radius decreasing steadily and its luminosity decreasing, roughly as R^2. In time, as the internal temperature continues to rise, ionization is completed and the opacity drops. The convective zone recedes from the centre and the star moves away from the Hayashi track toward higher effective temperatures. The increasing core temperatures cause nuclear reactions to start, slowly at first, far from thermal equilibrium, but gaining in intensity. This causes the stellar luminosity to reverse its trend and start rising. The evolution toward thermal equilibrium is complicated by the gradual ignition of different reactions of the hydrogen burning chains. This is illustrated in Figure 8.1 by the winding paths traced by stars of various masses in the ($\log L$, $\log T_{\text{eff}}$) diagram, obtained from detailed evolutionary calculations. The corresponding time intervals are listed in Table 8.1.

The relevant timescale throughout the protostellar stages is the relatively short Kelvin–Helmholtz (thermal) timescale given by equation (2.59). Stars in the pre-main-sequence evolutionary phase are hard to detect not only because they are scarce, this phase being relatively short, but also because they are still shrouded in the remains of the cloud out of which they were formed. The less massive among them, which evolve more slowly, appear as highly variable mass ejecting objects, known as *T Tauri* stars. They are surrounded by circumstellar discs, probable sites of planet formation, which are estimated to dissipate on timescales of up to 10^7 yr. An example of jets of material ejected by a young star hidden in a nebula of gas and dust is shown in Figure 8.2.

Only on the main sequence will the evolutionary timescale finally shift to the nuclear one and will stars become numerous. Contraction toward the main sequence takes up less than 1% of a star's life; in contrast, the star will spend about 80% of its life as a main-sequence star. For example, a star of $1 M_\odot$ spends 3×10^7 yr contracting prior to hydrogen ignition, in contrast to the 10^{10} yr it spends

8.1 The Hayashi zone and the pre-main-sequence phase

Table 8.1 Evolutionary lifetimes (years)

$M/M_\odot$	1–2	2–3	3–4	4–5
15	6.7(2)	2.6(4)	1.3(4)	6.0(3)
9	1.4(3)	7.8(4)	2.3(4)	1.8(4)
5	2.9(4)	2.8(5)	7.4(4)	6.8(4)
3	2.1(5)	1.0(6)	2.2(5)	2.8(5)
2.25	5.9(5)	2.2(6)	5.0(5)	6.7(5)
1.5	2.4(6)	6.3(6)	1.8(6)	3.0(6)
1.25	4.0(6)	1.0(7)	3.5(6)	1.0(7)
1.0	8.9(6)	1.6(7)	8.9(6)	1.6(7)
0.5	1.6(8)			

Note: powers of 10 are given in parentheses.

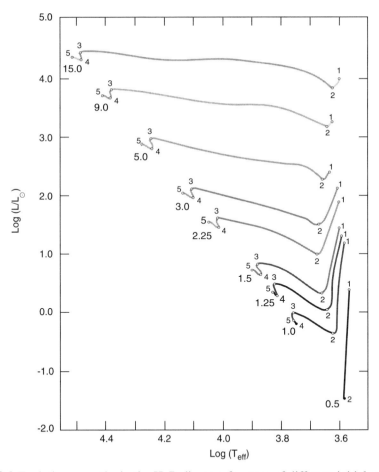

Figure 8.1 Evolutionary paths in the H–R diagram for stars of different initial mass (as marked) during the pre-main-sequence phase. The shade of segments is indicative of the time spent in each phase, ranging from less than 10^3 yr (light) to more than 10^7 yr (dark), as given in Table 8.1 [adapted from I. Iben Jr. (1965), *Astrophys. J.*, 141].

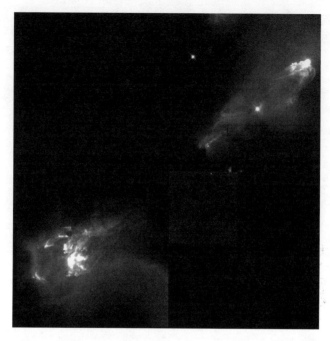

Figure 8.2 Jets of gas ejected by a young star. The jet spans more than 1 ly; the young star is hidden from view behind a dark cloud of dust [photograph by J. Hester, with NASA's Hubble Space Telescope].

burning hydrogen in the core. For initially more massive stars the timescales shrink significantly: thus for a $9M_\odot$ star the contraction phase takes only about 10^5 yr, and the main-sequence phase 2×10^7 yr.

8.2 The main-sequence phase

All stars undergo the main-sequence phase, characterized by core hydrogen burning. The major product is helium, but other important isotopes are synthesized during the main-sequence phase as well, such as nitrogen, resulting from proton capture on carbon nuclei in the stellar core, and the rarer isotopes ^{3}He and ^{13}C, produced in cooler regions, outside the core. In the course of this long phase there is ample time for the stellar configuration to achieve both hydrostatic and thermal equilibrium and "forget" its former structure and evolutionary phases. Thus the main sequence may be regarded as the starting point of stellar evolution; fortunately so, for the earlier stages of evolution are less well understood, partly perhaps because the guidance provided by observations is restricted by the short duration of these phases.

The nuclear energy generated in the hydrogen burning core is transported outward by radiation or by convection. In low-mass stars ($M \lesssim 0.3 M_\odot$) the main

8.2 The main-sequence phase

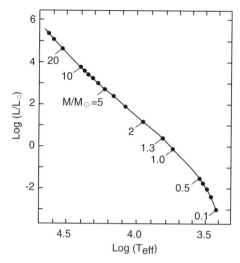

Figure 8.3 Correlation between luminosity and effective temperature obtained from model calculations of hydrogen burning stars of solar composition and various masses and the resulting main sequence in the H–R diagram [adapted from R. Kippenhahn & A. Weigert (1990), *Stellar Structure and Evolution*, Springer-Verlag].

means of energy transfer is convection, these stars being fully convective (except for the photosphere, of course), as we have anticipated in Section 6.6. In the H–R diagram they are found in the region where the Hayashi track meets the main sequence. More massive stars have smaller and smaller outer convective zones; in the Sun, for example, the convective zone extends over only about 2% of the solar mass below the photosphere. Stars more massive than the Sun, which burn hydrogen predominantly by the temperature-sensitive CNO cycle, develop convective cores, while the envelopes are in radiative equilibrium. The main sequence emerging from complex stellar model calculations is shown in Figure 8.3, with masses marked along it. The extent of convective zones is shown in Figure 8.4.

It is important to stress that the composition throughout a convective zone is uniform, as a result of continual mixing, even if the nuclear reaction rates are not. As a result, hydrogen burning products migrate to cooler regions of the star, where they could not have been found otherwise. And there they remain, even when the central convective zone shrinks or disappears altogether. Later on, when the stellar envelope will, eventually, become convective, its inner boundary overlapping the outer boundary of the formerly convective core, hydrogen burning products will make their way to the surface of the star, where they can be observed in the spectrum. This effect of consecutive, overlapping convecting zones, leading to the *dredge-up* of processed material to the surface, enables us to infer the occurrence of nuclear burning processes in the shielded stellar cores. Thus the detection of heavy elements in spectra of evolved stars and isotopic ratios different from those

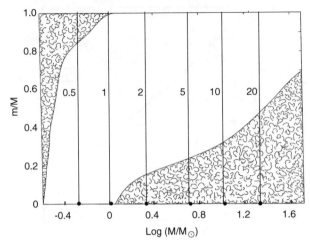

Figure 8.4 The extent of convective zones (shaded areas) in main-sequence star models as a function of the stellar mass [adapted from R. Kippenhahn & A. Weigert (1990), *Stellar Structure and Evolution*, Springer-Verlag].

of young stars constitutes another crucial test, as well as guide, to the theory of stellar evolution. Although there is ample observational evidence that indirectly validates the theory of nuclear energy generation in stellar interiors, great efforts are devoted to testing it directly. Experiments aiming to test the very hydrogen burning process taking place in the core of the Sun will be described in the next section.

One of the salient features of stellar evolution is mass loss. Stars lose mass at all evolutionary stages, including the main sequence. The rates of mass loss vary, however, over a very wide range. Thus, the mass loss rate on the lower main sequence is so slow as to have no discernible effect on the stellar mass. The solar wind, for example, removes mass from the Sun at a rate of $\sim 10^{-14} M_\odot$ yr^{-1}, which will amount to no more than 1/10 000 of the Sun's mass at the completion of its main-sequence phase.

Exercise 8.1: Assuming spherical symmetry, estimate the rate of mass loss from the Sun, if at Earth the measured velocity of the solar wind is ~ 400 km s^{-1} and the proton density in the wind is roughly 7 particles/cm^3.

As we go from low-mass to massive stars, the wind becomes more intense. As a result, although the main-sequence life span decreases rapidly with increasing stellar mass, the evolution pace of massive stars, which shed a considerable fraction of their mass by the wind, slows down compared with the evolutionary rate of these stars, had they conserved their mass. To illustrate this effect we define a

8.2 The main-sequence phase

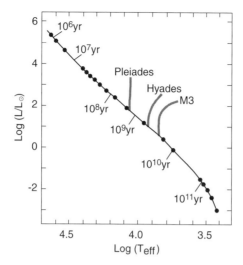

Figure 8.5 The main-sequence life span for stars of different masses marked along the main sequence in the H–R diagram (see Figure 8.3), which may be used to determine stellar cluster ages according to the "turning point" off the main sequence.

parameter α by

$$\alpha = \frac{\log(\tau_{MS}/\tau_{MS,\odot})}{\log(M/M_\odot)}.$$

We have seen in Chapter 7 that evolutionary timescales are determined by the stellar mass, as they are inversely proportional to the square (or a higher power) of the mass. Solar models are calibrated by requiring them to reproduce the solar radius and luminosity at the present age of the Sun, which is independently known from geologically based estimates of the age of the Earth. This calibration is used for computing the main-sequence phase of stars over a wide range of masses. The main-sequence lifetimes of stars within the initial mass range $0.1 M_\odot \leq M \leq 25 M_\odot$ are listed in Table 8.2 and are marked onto the main sequence in Figure 8.5, along with the corresponding stellar masses. The values of α are listed in the third column of Table 8.2: $|\alpha|$ is not constant; for relatively large M, it decreases with increasing M. We note the large span of main-sequence ages: at the lower end of the main sequence they exceed by far the age of the universe; by contrast, at the upper end, they become shorter than the thermal timescale of the Sun.

Consider now a stellar cluster, which is, essentially, a large group of stars born at the same time, more or less. The age of the cluster will show on its H–R diagram as the upper end, or *turning point* of the main sequence: stars within the cluster, with masses corresponding to main-sequence lifetimes shorter than the cluster's

Table 8.2 Main-sequence lifetimes

Mass ($M_\odot$)	Time (yr)	α
0.1	6×10^{12}	−2.8
0.5	7×10^{10}	−2.8
1.0	1×10^{10}	
1.25	4×10^{9}	−4.1
1.5	2×10^{9}	−4.0
3.0	2×10^{8}	−3.6
5.0	7×10^{7}	−3.1
9.0	2×10^{7}	−2.8
15	1×10^{7}	−2.6
25	6×10^{6}	−2.3

age, would have already left the main sequence toward the red giant branch. In other words, such stars would have consumed the hydrogen supply of their cores, the cores contracting toward the next burning stage (or toward becoming white dwarfs). Clearly, stars with main-sequence lifetimes longer than the cluster's age will still dwell on the main sequence of the H–R diagram. We are thus provided with a reliable tool (a *clock*) for measuring the age of star clusters, as illustrated in Figure 8.5, where superimposed on the *clock* are the H–R diagrams of different clusters. The oldest cluster provides a lower limit to the age of the galaxy within which it resides and to the age of the universe itself.

Note: The main sequence of stellar clusters serves not only as a time instrument but also as an instrument for distance determination. The H–R diagram obtained directly from observations has the measured apparent brightness, defined by equation (1.1), instead of the luminosity as ordinate. Since, on the logarithmic scale, this translates into a uniform vertical shift of magnitude $\log(4\pi d^2)$, matching such a diagram with the calibrated one enables the determination of the shift, and hence of the distance to the cluster. As the lower main sequence is the most populated region of the H–R diagram of any cluster, the matching procedure relies mainly on the main sequence, and thus this method of distance determination (which, in reality, is complicated by such factors as metallicity and interstellar absorption, cf. Chapter 10) is called *main sequence fitting*.

Stars of all masses partake in the main-sequence phase, but subsequent evolution differs for stars of different masses. In what follows we shall distinguish between stars whose main-sequence life span exceeds the present age of the universe (estimated as $1-2 \times 10^{10}$ yr) and stars that could have evolved off the main sequence, were they born early enough. Stellar models yield the upper mass limit for stars that are still on the main sequence (even if they are as old as the universe) at $0.7 M_\odot$. Due to their low surface temperature and thus reddish colour, these stars are also known as *red dwarfs*. Stars of mass $M > 0.7 M_\odot$ may be divided in turn into two subgroups according to their mass: those with initial masses below

9–10$M_\odot$, and the rest, with the former ending their lives as white dwarfs (after shedding a considerable fraction of the initial mass) and the latter undergoing supernova explosions. The former fall into *low-mass stars* ($0.7 \lesssim M \lesssim 2M_\odot$) and *intermediate-mass stars* ($2 \lesssim M \lesssim 9$–$10M_\odot$), and the latter are known as *massive stars*. The distinction between low-mass and intermediate-mass stars is based on the way of helium ignition in the core, that is, whether or not it occurs under degenerate conditions.

We now make a short digression from the course of stellar evolution to address the crucial issue of solar neutrinos.

8.3 Solar neutrinos

Since the mean free path of photons in stars is barely 1 cm, stellar cores, where nuclear reactions take place, cannot be directly observed. We infer the occurrence of nuclear reactions from the fact that stars shine and that their luminosities are well predicted by the theory of stellar evolution based on nuclear energy generation and also from the variety of surface abundances and isotopic ratios. However, a direct test of the theory would be possible by devising means of capturing neutrinos that are expelled in nuclear reactions. For them, the mean free path exceeds stellar dimensions by about 10 orders of magnitude. But if matter is so utterly transparent to these elusive particles, how can we expect to capture them? It turns out that a tiny fraction of the immense neutrino flux from the Sun (the nearest source) that sweeps the Earth *can* be captured by very ingenious experimental devices.

The energy generation process put to the test is thus the fusion of four protons into a helium nucleus, emitting two positrons and two neutrinos, and liberating thermal energy

$$4p \longrightarrow {}^4\text{He} + 2e^+ + 2\nu + Q,$$

where $Q \approx 25$ MeV after subtracting the average energy removed by the neutrinos. The number of neutrinos emanating from the Sun per second can be easily derived: $2L_\odot/Q \approx 2 \times 10^{38}$ s^{-1}. In the Sun, hydrogen burning proceeds mainly through the *p–p* reaction chain (described in Section 4.3), which is, in fact, the most common energy-generating process in stars. The chain, as we know, has three branches, involving three neutrino-emitting reactions. Due to the branching ratios of the *p–p* chain, the neutrinos emitted in each case have widely different fluxes, and also different energies, as given in Table 8.3. The branching ratios are directly related to the core temperature profile in the Sun (and vary from star to star). Now, the probability of absorption of a neutrino – small as it may be – increases with increasing energy; hence the easiest to detect would be the ^{8}B neutrinos, which have the highest energies.

Table 8.3 Properties of solar neutrinos

Source	Flux at Earth $(m^{-2}\,s^{-1})$	Energy (MeV)	Average (MeV)
$p + p \rightarrow {}^2D + e^+ + \nu$	6.0×10^{14}	≤ 0.42	0.263
${}^7Be + e^- \rightarrow {}^7Li + \nu$	4.9×10^{13}	0.86 (90%), 0.38 (10%)	0.80
${}^8B \rightarrow {}^8Be + e^+ + \nu$	5.7×10^{10}	≤ 15	7.2

Exercise 8.2: Using the data of Table 8.3, calculate the following: (a) the branching ratios of the p–p chain; (b) the neutrino luminosity of the Sun; and (c) the range of neutrino emission (particles per second) that would be expected, if the branching ratios of the p–p chain were *not* known.

Indeed, the ${}^8B$ neutrinos have been the main target of the first neutrino experiment, started in the early 1960s by Raymond Davis with the support of John Bahcall on the theoretical side, and still ongoing. The experiment is based on the capability of ${}^{37}Cl$ (a rare chlorine isotope), to absorb a high-energy neutrino and produce ${}^{37}Ar$, a radioactive isotope of argon,

$$\nu + {}^{37}Cl \longrightarrow e^- + {}^{37}Ar,$$

which subsequently decays, having a half-life of 35 days. Eventually, if the chlorine is exposed for a sufficiently long time, equilibrium is achieved between the production and destruction of argon. The equilibrium abundance of ${}^{37}Ar$ isotopes can be used to derive the flux of highly energetic neutrinos. As the threshold energy of the reaction is 0.8 MeV, these are mostly ${}^8B$ neutrinos; only a small fraction ($\sim 1/6$) result from ${}^7Be$. To grasp how formidable this experiment is, we should mention that it involves a huge tank containing about 600 tons of C_2Cl_4 fluid (an ordinary cleaning fluid) placed in the abandoned 1500-m deep Homestake gold mine in South Dakota. (The experiment has to be conducted deep underground in order to avoid background noise caused by cosmic ray particles.) The equilibrium abundance of argon atoms is no more than a few tens among the $\sim 2 \times 10^{30}$ chlorine atoms in the tank! Nevertheless, the radioactive argon atoms *can* be extracted and, by means of a Geiger counter, counted. The inferred neutrino flux being about a factor of 2 to 3 lower than predicted by solar models for the same energy range constitutes what has been known for many years as "the solar neutrino problem," and it is suspected of requiring new physics for its solution.

Another experiment aiming at the ${}^8B$ neutrinos is the more recent Kamiokande II, and its successor, the Super Kamiokande. The latter consists of a big tank – 40 m in diameter and 40 m high – containing 50 000 m^3 of very pure water, of which about half is used for the experiment itself, with the other half surrounding

8.3 Solar neutrinos

and shielding it. The tank is placed in the Kamioka Mozumi mine in Japan, at a depth of 2700 m. Its walls are lined with very sensitive light detectors capable of detecting single photons. (The earlier version used a smaller amount of water and yielded somewhat less accurate results.) The experiment is based on neutrino-electron scattering reactions,

$$\nu + e^- \longrightarrow \nu' + e^{-\prime},$$

which produce electrons moving with a speed that surpasses the speed of light in water (but is less than the speed of light in vacuum). Such electrons radiate energy, known as Cherenkov light, an effect that resembles a shock wave produced by an airplane moving at supersonic speed. This light hits the detectors on the tank walls. The threshold energy of this experiment is close to 7 MeV, and hence it is only sensitive to the more energetic among ^{8}B neutrinos. The number of events per day is less than 20. In principle, if one knows the neutrino flux detected by the Kamiokande experiment, one can predict from the energy distribution of the ^{8}B neutrinos the flux that should be detected by Davis's chlorine experiment. This is found to be higher than the flux actually detected, which further complicates "the solar neutrino problem." The remarkable achievement of the Kamiokande experiment is to have established that the detected neutrinos do indeed come from the Sun. The observed directions of the scattered electrons, which recoil in the direction of the scattering neutrinos, are found to trace out accurately the position of the Sun in the sky.

The disadvantage of both the chlorine and the water-Cherenkov experiments is that they test a rather insignificant branch of the p–p chain. The bulk of solar neutrinos are the low-energy ones produced by the fusion of two protons into deuterium. As it turns out, such neutrinos can interact with gallium,

$$\nu + {}^{71}\text{Ga} \longrightarrow e^- + {}^{71}\text{Ge},$$

the threshold energy being only 0.23 MeV, to produce radioactive germanium, which has a half-life of 11.4 days and decays back to gallium. Two experiments based on this reaction are operating: one named SAGE – a Russian (formerly Soviet)-American collaboration – in an underground excavation in the Caucasus region of Russia, and another named GALLEX – a primarily European collaboration – in an underground laboratory in Gran Sasso, Italy. SAGE uses 60 tons of metallic gallium (more than the amount produced worldwide in a year!); GALLEX uses half this amount in an aqueous solution. Similar to the method used in the chlorine experiment, the method of detecting the neutrinos is to collect and count the radioactive atoms in the target. More than half the neutrinos detected in these experiments come from the $p + p$ reaction. The comparison with theoretical predictions is significantly improved. The experimental results come within ∼65% of the solar model predictions, and the discrepancy is diminishing as more data accumulate and more refined effects are included in these models (such as diffusion, improved methods of dealing with convection, better opacities).

In conclusion, solar neutrinos have been observed in five different experiments, with roughly the expected fluxes (allowing for the apparent deficiency of ^{8}B neutrinos) and energies. Moreover, it has been unequivocally confirmed that their source is the Sun. We may safely claim that the main goal of the neutrino detection experiments has been attained. Further experiments, even more sophisticated and sensitive, are currently being planned and mounted (known by the names of SNO and BOREXINO). It is quite possible that only a few years from now the solar neutrino problem will have been finally solved.

8.4 The red giant phase

As the main-sequence phase advances, a hydrogen-depleted core grows gradually in mass. Hydrogen burning proceeds in a shell surrounding the core, which separates it from the envelope. Since the core is now devoid of energy sources, the heat flow through it falls to zero [from equation (5.4), $F = \int q\,dm \to 0$] and with it the temperature gradient decreases [equation (5.3)]. Thus, as the burning shell moves outward, the core becomes isothermal while its mass increases. We shall now show that an isothermal core of ideal gas cannot have an arbitrarily large mass: given the stellar mass M, an upper limit exists for the core mass M_c, beyond which the pressure within the core is incapable of sustaining the weight of the overlying envelope. M. Schönberg and Chandrasekhar were the first to point out and derive the limiting mass from model calculations in 1942, and this type of dynamical instability is therefore known as the Schönberg–Chandrasekhar instability. It can be easily understood on the basis of the virial theorem (Section 2.4) – as William McCrea showed 15 years later, although in a completely different context (McCrea was studying star formation by gravitational collapse).

Denoting the core radius by R_c, its volume by V_c, the mean molecular weight within it by μ_c, and the temperature by T_c, we have from equation (2.24)

$$\int_0^{V_c} P\,dV = P_s V_c + \tfrac{1}{3}\alpha \frac{GM_c^2}{R_c}, \qquad (8.17)$$

where P_s is the pressure at the core's boundary. Now, for an ideal isothermal gas,

$$\int_0^{V_c} P\,dV = \frac{\mathcal{R}}{\mu_c} T_c \int \rho\,dV = \frac{\mathcal{R}}{\mu_c} T_c M_c. \qquad (8.18)$$

8.4 The red giant phase

Substituting relation (8.18) and $V_c = 4\pi R_c^3/3$ into equation (8.17), we obtain

$$P_s(R_c) = \frac{3}{4\pi} \frac{\mathcal{R} T_c}{\mu_c} \frac{M_c}{R_c^3} - \frac{\alpha G}{4\pi} \frac{M_c^2}{R_c^4}. \tag{8.19}$$

For a given core mass, the pressure at the core boundary increases with the core radius from $P_s = 0$ at

$$R_0 = \frac{\alpha G}{3\mathcal{R}} \frac{M_c \mu_c}{T_c}, \tag{8.20}$$

to a maximum value $P_{s,\text{max}}$ at

$$R_1 = \frac{4\alpha G}{9\mathcal{R}} \frac{M_c \mu_c}{T_c}, \tag{8.21}$$

obtained by setting $dP_s/dR_c = 0$. A core of radius $R_c < R_0$ would collapse under its own gravitation, without any external pressure. For a core of radius $R_c > R_1$, the pressure at its boundary would be smaller than $P_{s,\text{max}}$. Thus the maximal pressure that can be attained at the core boundary as a function of the core mass is

$$P_{s,\text{max}}(M_c) = \text{constant} \, \frac{T_c^4}{M_c^2 \mu_c^4}. \tag{8.22}$$

This pressure, exerted by the gas within the core, must balance the pressure P_{env} exerted by the envelope. To estimate the latter, we may assume the core to be a point mass ($R_c \ll R$) and make use of inequality (2.18) obtained in Section 2.3: $P_{\text{env}} > GM^2/8\pi R^4$. Obviously, if $P_{s,\text{max}} < GM^2/8\pi R^4$, no equilibrium configuration would be possible. Hence the stability condition for an isothermal core is

$$P_{s,\text{max}}(M_c) = \text{constant} \, \frac{T_c^4}{M_c^2 \mu_c^4} \geq \frac{GM^2}{8\pi R^4}. \tag{8.23}$$

Still regarding the core as a point mass, we may use homology relation (7.33) of Section 7.4,

$$T_c \propto \frac{\mu_{\text{env}}}{\mathcal{R}} \frac{GM}{R},$$

to eliminate T_c and R in condition (8.23). The stability condition thus becomes

$$\frac{M_c}{M} \lesssim \text{constant} \left(\frac{\mu_{\text{env}}}{\mu_c}\right)^2. \tag{8.24}$$

Schönberg and Chandrasekhar arrived at this result with the dimensionless constant of 0.37. Assuming a solar composition for the envelope and a mostly helium composition for the core, we have by equations (3.29), (3.26), and (3.18) $\mu_{\mathrm{env}} \simeq 0.6$ and $\mu_c \simeq 1$, leading to $M_c/M \lesssim 0.13$. When the mass of the hydrogen-depleted core reaches this limit, the core starts contracting rapidly.

Main-sequence stars more massive than about $2M_\odot$ have homogeneous convective cores surpassing the critical limit, as shown in Figure 8.4. Once hydrogen is exhausted in such a core, energy generation subsides, convection is quenched, and the core becomes isothermal. Since its mass is already greater than the Schönberg–Chandrasekhar limit, the dynamically unstable core starts collapsing. In time, it acquires the temperature gradient necessary for balancing gravity. The temperature gradient causes loss of heat, and hence core contraction and the increase in temperature that goes with it continue, but on a thermal (Kelvin–Helmholtz) timescale.

When hydrogen burning in the core ceases, thermal equilibrium is destroyed and for a brief period of time the stellar energy decreases ($L > L_{\mathrm{nuc}}$). However, as hydrogen burning shifts from the core to a shell surrounding it, and as the temperature in this shell rises with the rising core temperature, the nuclear energy generation rate soon increases again. But since hydrogen is burnt by the CNO cycle, whose rate varies as a very high power of the temperature (see Section 4.4), the energy production rate is accelerated beyond thermal equilibrium, and during most of the core contraction phase the stellar energy increases ($L_{\mathrm{nuc}} > L$). This is illustrated in the top-left panel of Figure 8.6, where the change with time of the total energy of a $7M_\odot$ star model is plotted, beginning at the end of the main sequence and ending on the red giant branch. Core contraction is thus necessarily accompanied by expansion of the envelope (cf. Section 7.5), as illustrated in the bottom-left panel of Figure 8.6, and the star becomes a red giant, *moving to the right* in the H–R diagram, as shown in the bottom-right panel of the same figure. Overall, the transition from a main-sequence to a red giant configuration is characteristically of short duration, and hence the probability of detecting stars undergoing this transition is vanishingly small. This is the reason for the conspicuous gap between the main sequence and the red giant branch in the H–R diagram, known as the *Hertzsprung gap*.

Note: The question "How does a star become a red giant?" constitutes a long-standing puzzle. But the puzzle is connected not so much with the physics of red giants as with our perception of *understanding* a phenomenon. We may claim to understand a physical process in the following cases: (a) if we can lay down the physical principles governing it; (b) if we can write down the equations describing it and solve them; (c) if we can explain the process in simple terms, step by step. Of course, if all three conditions are fulfilled, the process may be considered well understood. But in fact, condition (b) alone suffices. This is the case with red giants: all numerical computations of the evolutionary phase following hydrogen

8.4 The red giant phase

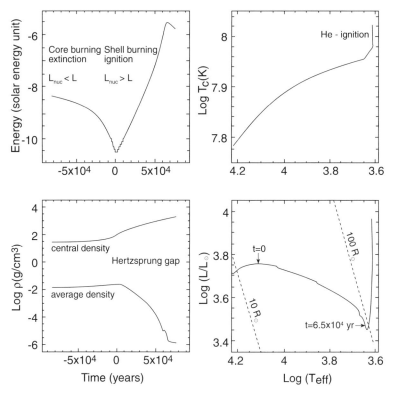

Figure 8.6 Evolution of an intermediate mass star ($7M_\odot$) during the crossing of the *Hertzsprung gap*: *top left*: total energy as a function of time (the time is arbitrarily set to zero at the onset of core contraction); *bottom left*: central density and average density ($3M/4\pi R^3$) as a function of time; *bottom right*: evolutionary track in the H–R diagram (where lines of equal radius are marked); *top right*: changing of central temperature with effective temperature.

exhaustion in the core obtain red giant configurations as solutions of the stellar evolution equations. Moreover, the simple explanation offered in Section 7.5 points out the virial theorem as the basic principle involved, given the contraction of the core – thus satisfying condition (a). Nevertheless, we feel uncomfortable in accepting these solutions so long as condition (c) is not satisfied. We would like to be able to identify the precise mechanism that drives a star to become a red giant. However, this last condition is not always considered imperative for understanding a physical process. For example, we understand and explain the outcome of a collision of two rigid balls on a smooth surface in terms of conservation of momentum and energy, without bothering about the exact manner in which momentum is transferred from one ball to another during their brief contact. And yet we still worry about red giants ...

As the helium core grows in mass by hydrogen burning in the shell outside it, it continues to contract, liberating gravitational energy. Consequently, the temperatures in the core and shell go on rising (see the top-right panel of Figure 8.6), accelerating further the rate of hydrogen burning and core growth. Finally, thermal equilibrium is restored and the luminosity, which is proportional to the rate of core growth, increases. The need to transfer an increasing energy flux on the one hand, and the increasing opacity of the cool envelope on the other hand, cause the envelope to become convectively unstable. Hence red giants develop convective envelopes, extending from just outside the hydrogen burning shell all the way to the surface. The base of the convective envelopes reaches layers where nuclear processes have taken place earlier and thus hydrogen burning ashes make their way to the surface. This is the first occurrence of "dredge-up" (explained in Section 8.2) that is observationally detected. In the ($\log L$, $\log T_{\text{eff}}$) plane these stars are said to *climb* up the slanted red giant branch (very close to their Hayashi tracks) toward higher luminosities and slightly lower effective temperatures. The red giant branch in the H–R diagram roughly coincides with the boundary of the Hayashi forbidden zone. Eventually, the core temperature becomes sufficiently high for helium to ignite.

The Schönberg–Chandrasekhar instability applies, however, only to ideal gases. A cold and dense gas, in which the degenerate electrons supply most of the pressure, is capable of building up a sufficient degeneracy pressure to support the weight of the envelope, even in a relatively massive core. The appropriate conditions for electron degeneracy,

$$P_{s,\max}(M_c) \lesssim K_1 \left(\frac{M_c}{\frac{4\pi}{3} R_c^3} \right)^{5/3},$$

using equation (3.35), are found to develop in the helium cores of stars with masses below about $2M_\odot$. The core contraction phase of these stars is slow and gradual. The temperature rises throughout the contracting core and the burning shell outside it. As a result, the nuclear energy generation rate increases and, with it, the stellar luminosity. At the same time, the envelope expands and the temperature decreases throughout it, as well as at the stellar surface. The star assumes *gradually* the appearance of a red giant. Indeed, the ascent toward the red giant branch is clearly seen at the lower part of the main sequence, in particular in H–R diagrams of old globular clusters. However, these low-mass stars, evolving quietly toward higher core temperatures, are bound to encounter a different type of instability. Heating of the core as a result of contraction is impeded by neutrino emission, which acts as an energy sink. Hence the core material becomes strongly degenerate before helium burning sets in. We have seen in Section 6.2 that nuclear burning in degenerate material is thermally unstable, leading to a *runaway*. Thus in these relatively low-mass stars, when the temperature finally reaches the helium ignition threshold, helium ignites in an explosion, known as the *helium*

8.5 Helium burning in the core

flash. This occurs when the core mass has grown to about $0.5M_\odot$, regardless of the total stellar mass. During a few seconds, the temperature rises steeply at almost constant density, the local nuclear power reaching 10^{11} $L_\odot$ (roughly, the luminosity of an entire galaxy). Nevertheless, an outsider would not be aware of the intense central explosion, which is almost entirely quenched by the energy-absorbing stellar envelope. Thus, there is no apparent clue in the H–R diagram to the helium flash. Soon, the core temperature becomes sufficiently high for the degeneracy to be lifted, the core expands, and helium burning becomes stable.

A fraction of the red giants, however, do not attain helium ignition. This is due to the effect of mass loss that characterizes red giants. During the red giant phase, when the stellar envelope is considerably less bound gravitationally than in the main-sequence phase, the stellar wind intensifies. Hence low-mass red giants lose their small envelopes before the core has a chance to reach sufficiently high temperatures for helium ignition. The degenerate helium cores continue their contraction, leaving the red giant branch to become helium white dwarfs.

Exercise 8.3: Assume a star of mass M and radius R has a core of mass M_1 and radius R_1. Let the density distribution be given by

$$\rho = \rho_c - (\rho_c - \rho_1)\left(\frac{r}{R_1}\right)^2 \quad \text{for} \quad 0 \leq r \leq R_1$$

$$\rho = \rho_1 \frac{\left(\frac{R_1}{r}\right)^3 - \left(\frac{R_1}{R}\right)^3}{1 - \left(\frac{R_1}{R}\right)^3} \quad \text{for} \quad R_1 \leq r \leq R,$$

where ρ_c is the central density and $\rho_1 = \rho(R_1)$. Find the dependence of the ratio R/R_1 on $x_1 \equiv \rho_c/\rho_1$ and $y_1 \equiv M/M_1$. Calculate the ratio for $x_1 = 10$ and $y_1 = 7.5$ [consistent with condition (8.24)].

8.5 Helium burning in the core

The phase of stable helium burning in the stellar core is significantly shorter than the main-sequence phase of core hydrogen burning. The reason is twofold: first, the fusion of helium – into carbon and oxygen – supplies only about one tenth of the energy per unit mass supplied by hydrogen fusion (as we have seen in Chapter 4), and secondly, the stellar luminosity is higher by more than an order of magnitude compared with the main-sequence luminosity of the same star. In fact, helium burning would have been still shorter, were it not for the additional energy source provided by hydrogen burning in the shell outside the core.

In low-mass stars (0.7–$2M_\odot$), which undergo the helium flash, the subsequent rapid expansion of the core has an effect on the star's structure similar to the contraction of the core at the end of the main sequence, only in reverse. As the core expands and cools, the envelope contracts and its temperature rises to some extent. As a result of core expansion and cooling, the temperature in the hydrogen burning shell decreases and the nuclear energy supply diminishes. This, combined with the diminished stellar radius, causes the luminosity to drop and the star is said to *descend* from the red giant branch. Since the effective temperature has increased, the star *moves to the left* in the H–R diagram. The locus of low-mass core helium burning stars in the H–R diagram forms the *horizontal branch*, a roughly horizontal strip stretching between the main sequence and the red giant branch, corresponding to luminosities of the order of 50–$100L_\odot$. There they dwell for about 10^8 yr. All these stars have equally massive cores at the end of the red giant phase; hence their different positions along the horizontal branch must be determined by another factor. For stars of similar Z (heavy element content), this factor is found to be the envelope mass, a function of the initial stellar mass and the rate of mass loss up to this stage, itself possibly a function of the rotation rate of the star. The highest envelope masses are found at the red (low effective temperature) end of the branch, where the hydrogen shell contributes most of the energy and the convective envelope's structure is similar to that of red giants. Proceeding toward the blue end, we find smaller envelope masses and weaker hydrogen burning shells. The envelopes are now radiative rather than convective. Stars in this region of the horizontal branch are found to go through a phase of dynamical instability in their envelopes, in the regions of hydrogen and helium ionization (see Section 6.4). This instability manifests itself by pulsations, causing a cyclic variability of the luminosity with periods of a few hours. Such pulsating stars on the horizontal branch are indeed observed; they are known as *RR Lyrae variables*. At the blue end of the branch the hydrogen-rich envelopes are small – in both mass and radius – and inert.

Intermediate-mass stars (2–$10M_\odot$) ignite helium quietly when the central temperature reaches 10^8 K. Subsequently, the rate of energy supply by the helium burning core steadily increases, while the rate of energy supply by the hydrogen burning shell decreases. As the temperature in the burning shell at the base of the envelope drops, the envelope cools too and, eventually, it starts contracting; this occurs when the contributions from the two energy sources become roughly equal. At this point the stars leave the red giant branch in the H–R diagram by looping toward higher effective temperatures; the higher the mass, the more extended the loop. As luminosity increases with mass, these stars form a *helium main sequence*, with a slope similar to that of the (hydrogen) main sequence, but closer to the red giant branch. In fact, observationally, the helium main sequence

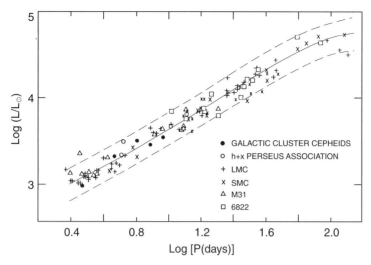

Figure 8.7 The period-luminosity correlation for Cepheids derived from observations [from A. Sandage & G. A. Tammann (1968), *Astrophys. J.*, 151].

is hardly discernible from the thick red giant branch. Intermediate-mass helium burning stars, too, go through a phase of envelope instability resulting in pulsations, but the pulsation periods are longer, ranging from days to several months. Such pulsating luminous stars are known as *Cepheid variables*, or simply *Cepheids*. Their importance to astronomy warrants another digression from the pursuit of stellar evolution.

It turns out that a well-defined correlation exists between the (average) luminosity of a Cepheid star L_{Ceph} and its pulsation period P_{Ceph}, as shown in Figure 8.7. The correlation emerges from observations of Cepheids with well-known distances, for which accurate luminosities can be derived, and thus $L_{\text{Ceph}}(P_{\text{Ceph}})$ is established. Imagine now that a pulsating star is detected in a distant cluster of stars or a distant galaxy, with a period P_{obs} characteristic of Cepheids. If the star is identified as a Cepheid (based also on its spectral characteristics), its apparent brightness I_{obs} and its pulsation period can be used to derive its distance d, which is also the distance to the cluster or galaxy within which it resides

$$d = \left[\frac{L_{\text{Ceph}}(P_{\text{obs}})}{4\pi I_{\text{obs}}} \right]^{1/2}. \tag{8.25}$$

Cepheids constitute what are called in astronomy *standard candles* – the most accurate and reliable among them. The period–luminosity relationship was first discovered in 1912 by Henrietta Leavitt, for the Cepheids in the nearby galaxy called the *Small Magellanic Cloud* (or SMC), and these stars immediately rose to

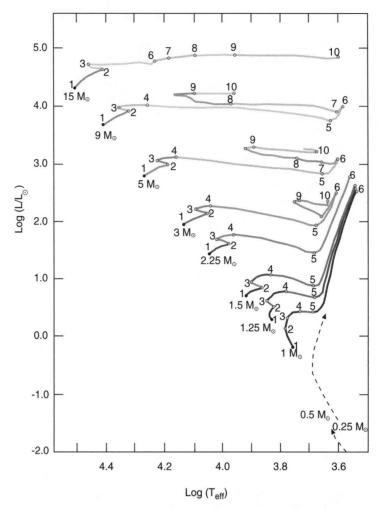

Figure 8.8 Evolutionary paths in the H–R diagram for stars of different initial masses (as marked) up to the stage of helium burning in a shell. The shade of the segments is indicative of the time spent in each phase, ranging from less than 10^5 yr (light) to more than 10^9 yr (dark), as given in Table 8.4. The different phases, indicated by numbers, are: 1–2, main sequence; 2–3, overall contraction; 3–5, H burning in thick shell; 5–6, shell narrowing; 6–7, red giant branch; 7–10, core helium burning; 8–9, envelope contraction [adapted from I. Iben Jr. (1967), *Ann. Rev. Astron. Astrophys.*, 5].

fame. A year after the discovery, the period-luminosity relation had already been used by Hertzsprung and other noteworthy astronomers to determine distances to galaxies.

Note: Within our own Galaxy the relative brightness of stars in a given volume is largely affected by their different distances. Stars of a distant galaxy, however, are all equally distant from an observer on Earth, because the distance to a galaxy

8.6 Thermal pulses and the asymptotic giant branch

Table 8.4 Evolutionary lifetimes (years)

$M/M_\odot$	1–2	2–3	3–4	4–5	5–6	6–7	7–8	8–9	9–10
15	1.0(7)	2.3(5)	←	7.6(4)	→	7.2(5)	6.2(5)	1.9(5)	3.5(4)
9	2.1(7)	6.1(5)	9.1(4)	1.5(5)	6.6(4)	4.9(5)	9.5(4)	3.3(6)	1.6(5)
5	6.5(7)	2.2(6)	1.4(6)	7.5(5)	4.9(5)	6.1(6)	1.0(6)	9.0(6)	9.3(5)
3	2.2(8)	1.0(7)	1.0(7)	4.5(6)	4.2(6)	←	6.6(7)	→	6.0(6)
2.25	4.8(8)	1.6(7)	3.7(7)	1.3(7)	3.8(7)				
1.5	1.6(9)	8.1(7)	3.5(8)	1.0(8)	>2(8)				
1.25	2.8(9)	1.8(8)	1.0(9)	1.5(8)	>4(8)				
1.0	7.0(9)	2.0(9)	1.2(9)	1.6(9)	>1(9)				

Note: Powers of 10 are given in parentheses.

is far larger than its size. Consequently the ratio of apparent brightnesses is equal to the ratio of intrinsic luminosities for these stars. Hence statistical analyses are far more reliable for stars in the nearby galaxies, such as the Magellanic Clouds.

Due to its high temperature sensitivity, helium burning occurs in a convective core (as does hydrogen burning by means of the CNO cycle). This is the inner part of the larger helium core, which grows as a result of hydrogen burning in the shell surrounding it. So long as the inner core is convective, its composition is constantly mixed and turns gradually from helium to carbon and oxygen, although helium burning, which varies with temperature as $\sim T^{40}$, is confined to its very centre. When the inner convective core becomes depleted of helium and hence burning within it subsides, convection is quenched as well. The star consists of a carbon-oxygen core, surrounded by a helium layer – remnant of the original helium core – which, in turn, is separated from the hydrogen-rich envelope by the hydrogen burning shell. Helium burning continues in a shell at the C-O core boundary. Evolutionary paths in the H–R diagram up to helium burning in a shell are traced in Figure 8.8 for stars of different masses, and the characteristic time intervals are listed in Table 8.4. A comparison between theory and observations is made possible by varying the shade of the paths in Figure 8.8 in proportion to the length of time spent – and hence to the number of stars expected to be observed – in each phase. Among core helium burning stars, those observed are predominantly the low-mass ones populating the horizontal branch.

8.6 Thermal pulses and the asymptotic giant branch

The C-O core's evolutionary course and its consequences are similar to the helium core's. Devoid of energy sources, the core contracts and heats up; as a result, the

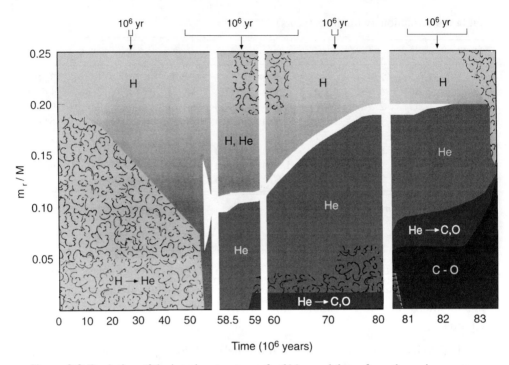

Figure 8.9 Evolution of the interior structure of a $6M_\odot$ model star from the main sequence to the AGB phase. Dark areas indicate nuclear burning and shaded ones indicate convective zones. Note the occasional changes in timescale.

envelope expands and cools, and convection sets in again throughout it. As the inner boundary of the convective envelope overlaps the earlier outer boundary of the now extinguished hydrogen burning shell, processed material, mainly helium and nitrogen, is once more dredged up and mixed into the envelope. The signature of these elements appears in the star's spectrum, again bearing witness to the processes taking place in its deep interior. The expanding star becomes redder and resumes its climb on the giant branch in the H–R diagram, which has been interrupted by the core helium burning episode. This part of the giant branch, populated by stars with C-O cores, is called *the asymptotic giant branch*; it is an extension of the red giant branch toward higher luminosities and lower effective temperatures at the boundary of the Hayashi forbidden zone. Hence stars in this phase of evolution (known as AGB stars) are even bigger than the former red giants – they are now becoming *supergiants*. Cooling of the layers above the C-O core extinguishes the hydrogen burning shell temporarily; it will reignite later on, after envelope expansion will come to a halt. Contraction of the core raises the density up to the point when electrons become degenerate, and since degenerate matter is an efficient heat conductor, the core becomes isothermal. An illustration of the internal evolution of a $6M_\odot$ star from the main sequence up to the onset of the asymptotic giant phase is given in Figure 8.9: changing burning zones, convective zones, and boundaries between regions of different composition are marked. The remarkably different evolutionary timescales are particularly noteworthy.

There are three outstanding characteristics for AGB stars:

1. Nuclear burning takes place in two shells – a thermally unstable configuration – leading to a long series of *thermal pulses*.
2. The luminosity is uniquely determined by the core mass, independently of the total mass of the star.
3. A strong stellar wind develops as a result of the high radiation pressure in the envelope, the star thus losing a significant fraction of its mass.

We shall now address each characteristic in more detail. The two burning shells that supply energy during the asymptotic giant phase are separated by a helium layer. The external shell, at the bottom of the hydrogen-rich envelope, burns hydrogen, thus increasing the helium layer's mass. The internal shell, on top the C-O core, burns helium, thus eating into the helium layer and building up the C-O core. In principle, a steady state could be achieved, with the two burning fronts advancing outward at the same rate. However, the great differences between the two nuclear burning processes do not allow such a steady state to develop. As it happens, the two shells do not supply energy concomitantly, but in turn, in a cyclic process, and the mass of the helium layer separating them changes periodically.

During most of the cycle's duration hydrogen is burnt in the external shell, while the inner shell is extinct. As a result, the helium layer separating the shells grows in mass. With no energy supply, this layer contracts and heats up until the temperature at its base becomes sufficiently high for helium to ignite. Helium burning in this thin shell is thermally unstable, as explained in Section 6.2; it resembles the helium flash that takes place in the electron-degenerate cores of low-mass stars at the tip of the giant branch. At the peak of the short-lived flash the nuclear energy generation rate reaches $10^8 L_\odot$. The energy is absorbed by the overlying layers, which expand and cool. As these layers contain the hydrogen burning shell, the rate of hydrogen burning quickly declines. During an ensuing short period of time, the helium burning front advances through the helium shell, turning helium into carbon and oxygen, until it catches up with the now extinct hydrogen shell.

The high temperatures attained in the helium burning shell lead to a chain of reactions that produces neutrons. Capture of these neutrons by traces of heavy elements that are present in the shell leads to the creation of trans-iron isotopes by the *s*-process explained in Section 4.8.

The proximity of the hot helium burning front causes the hydrogen to reignite. Because of its lesser sensitivity to temperature, hydrogen burning in a shell is stable. The temperature and density adjust into thermal equilibrium. At the same

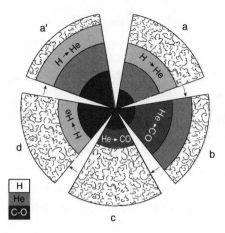

Figure 8.10 Sketch of the progress of a thermal pulse cycle through its different stages (not in scale). Hydrogen is burning during stages a and d, while helium is burning during stages b and c. When, in stage c, the outer convective zone extends inward beyond the helium shell burning boundary, hydrogen and helium burning products are mixed into the envelope and dredged up to the surface. Stage a' is the same as a, except that the carbon-oxygen core has grown at the expense of the envelope.

time, helium burning is quenched as a result of the relatively low temperature now prevailing in the hydrogen burning shell and its vicinity. Thus a new cycle begins. The evolution throughout a thermal pulse cycle, known also as a *shell flash*, is shown schematically in Figure 8.10. Particularly noteworthy is the dredge-up of processed material into the convective envelope by the moving inner boundary of the convective zone.

Although thermal pulses entail periodic changes of the stellar luminosity, these cannot be observed because the periods vary between hundreds to thousands of years. The lasting result of each cycle is the growth of the C-O core. This brings us to the second characteristic of AGB stars.

Evolutionary calculations show that the luminosity of an asymptotic giant branch star is quite accurately represented, for core mass $M_c > 0.5\ M_\odot$ (recalling that the star is now well past the onset of helium burning), by the following relation:

$$\frac{L}{L_\odot} = 6 \times 10^4 \left(\frac{M_c}{M_\odot} - 0.5\right), \tag{8.26}$$

as was first pointed out by Bohdan Paczyński in 1971. (We note that this luminosity is of the order of the Eddington luminosity $L/L_\odot = 3.2 \times 10^4 M/M_\odot$; see Section 5.5.) Thus stars of the same core mass are found at the same height

8.6 Thermal pulses and the asymptotic giant branch

on the asymptotic giant branch in the H–R diagram, regardless of their envelope mass. Stars reach the asymptotic giant branch at different points, depending on the mass of the core at the end of the central helium burning stage. They climb up the branch, as the core continues to grow during the thermal pulses stage. At the same time the envelope mass decreases, not only at the expense of core growth, but mainly because of mass loss at the surface. Hence the point at which a star leaves the asymptotic giant branch is determined by the mass of the envelope at the end of core helium burning and by the intensity of the stellar wind. This brings us to the third characteristic of AGB stars.

The stellar wind phenomenon is not yet well understood theoretically, except for the recognition that mass loss is driven by the increasingly dominant radiation pressure, as the stellar luminosity approaches the Eddington limit. Radiation pressure is capable of accelerating material out of the stellar gravitational potential well, notwithstanding the overall state of hydrostatic equilibrium, because material particles vary widely in their ability to absorb radiation. While the interaction of a particle with a gravitational field depends solely on the particle's mass, its interaction with a radiation field depends on its composition, structure, size, and density, as well as on the radiation wavelength. Thus, if in the outer layers of a star there are such particles that are exceptionally absorbent at the leading wavelength of the photons – as determined by the temperature – then for these particles the radiation pressure might just overcome gravity. The result would be an outward acceleration leading to a mass outflow of such particles, and others entrained by them. At a mass loss rate $\dot{M}$ driven by radiation pressure, the mass $\dot{M}\delta t$ ejected during a time interval δt acquires escape velocity by absorbing a fraction, say ϕ', of the momentum carried by the radiation $(L/c)\delta t$. Consequently,

$$\dot{M}\delta t\, v_{\text{esc}} = \phi' \frac{L}{c} \delta t,$$

and substituting $v_{\text{esc}}^2 = 2GM/R$, and $\phi = \phi'/2$, we may write

$$\dot{M} = \phi \frac{v_{\text{esc}}}{c} \frac{LR}{GM}. \tag{8.27}$$

Now, the outer layers of giant and supergiant stars are sufficiently cool for atoms to coalesce into molecules and molecules into tiny dust particles. It is these particles that are accelerated by the radiation pressure. However, the nature of such particles and the interactions involved are extremely difficult to calculate and the resulting mass loss rate (or parameter ϕ) difficult to assess. At this point, the stellar evolution theory has to rely on observations in order to continue its pursuit of the changing structure of stars.

> **Exercise 8.4:** (a) Estimate the mass loss timescale, τ_{m-1}, and compare it with the thermal timescale of a star. (b) Show that the rate of energy supply required for mass loss at a rate $\dot{M}$ is a very small fraction of L. (c) Find the relation between the mass loss timescale and the nuclear timescale of the star and show that, usually, $\tau_{m-1} < \tau_{\text{nuc}}$.
>
> **Exercise 8.5:** Assuming that the mass loss rate may be parametrized as in equation (8.27): $\dot{M} \propto LR/GM$, show that for main-sequence stars $\dot{M} \propto L^{\alpha}$ and evaluate α.

Observations of red giants and supergiants reveal that these stars lose mass at rates ranging from 10^{-9} to 10^{-4} $M_\odot$ per year. Mass loss is generally classified into two types of winds.

1. A stellar wind that may be described by an empirical formula from Dieter Reimers, linking the stellar mass, radius, and luminosity, by a relation of the form of relation (8.27), the constant being determined from observations over a wide range of stellar parameters:

$$\dot{M} \approx 10^{-13} \frac{L}{L_\odot} \frac{R}{R_\odot} \frac{M_\odot}{M} \quad M_\odot \text{ yr}^{-1}. \qquad (8.27')$$

   Typical wind rates are of the order of 10^{-6} $M_\odot$ yr^{-1}, which for characteristic M, L, and R values, imply $\phi \sim 1$ in equation (8.27), that is, a high efficiency in momentum transfer.
2. A *superwind*, essentially a stronger wind, leading to a concentration of the stellar ejecta in an observable shell surrounding the central star.

8.7 The superwind and the planetary nebula phase

The existence of the superwind is imposed by two different and independent observations: first, the high density within the observed shells formed by the stellar ejecta (a slow wind would have given rise to more diffuse shells), and secondly, the relative paucity of very bright stars on the asymptotic giant branch of H–R diagrams of statistically significant stellar samples.

The inference of a strong wind based on the second observation is not straightforward, but it can be understood in terms of simple arguments. The number of stars expected to reside on the asymptotic giant branch in the H–R diagram is proportional to the time spent by stars in the double-shell burning evolutionary phase. An upper limit to this time span may be obtained by considering only

8.7 The superwind and the planetary nebula phase

the evolution of the core. Essentially, AGB stars turn hydrogen into carbon and oxygen, growing C-O cores at the expense of envelope material. If Q is the amount of nuclear energy released in the process per unit mass, roughly 5×10^{14} J kg^{-1}, thermal equilibrium implies

$$\dot{M}_c = \frac{L}{Q} \approx 1.2 \times 10^{-11} \frac{L}{L_\odot} \, M_\odot \, \text{yr}^{-1}, \tag{8.28}$$

where L is the luminosity (averaged over a thermal pulse cycle). Substituting $L/L_\odot$ from equation (8.26), we obtain

$$\frac{dM_c}{M_c - 0.5 M_\odot} = 7.2 \times 10^{-7} \, \text{yr}^{-1} \, dt. \tag{8.29}$$

At the beginning of the asymptotic giant phase the core has some mass $M_{c,0}$ ($> 0.5 M_\odot$). Assuming that as a result of contraction the electrons become degenerate, the maximal mass the core could reach is the Chandrasekhar mass M_{Ch}. Integrating between $M_c = M_{c,0}$ and $M_c = M_{Ch}$, we obtain an upper limit to the duration of the asymptotic giant phase,

$$\tau_{AGB} < 1.4 \times 10^6 \ln \left(\frac{M_{Ch} - 0.5 M_\odot}{M_{c,0} - 0.5 M_\odot} \right) \, \text{yr}. \tag{8.30}$$

Evolutionary calculations show that a relation exists between the initial core mass and the initial mass of the star M_0, of the form

$$M_{c,0} \approx a + bM_0, \tag{8.31}$$

where a and b are constants. Hence τ_{AGB} is, essentially, a function of the initial stellar mass. Therefore, if the distribution of initial stellar masses is known, the number of stars on the asymptotic giant branch can be computed for any given population of stars. As it turns out, the expected number of AGB stars exceeds by far the actual number of observed AGB stars, with the discrepancy being as large as a factor of 10. This means that stars are prevented by some process from completing their sojourn on the asymptotic giant branch, while losing mass at the moderate rate dictated by Reimers's formula. This process is the superwind, which consumes the envelope mass before the core has grown to its maximal possible size. In fact, mass loss must be so intense as to allow the core to grow by only about $0.1 M_\odot$ while the entire envelope is ejected. It should be mentioned that, in addition to the indirect indications, the hypothesis of a superwind is confirmed by observations of stars which eject mass at rates of the order of $10^{-4} M_\odot$ yr^{-1}. In some AGB stars, those believed to descend from relatively low-mass progenitors, the high mass loss rate is associated with a pulsation instability in the envelope, similar to that of RR Lyrae stars and Cepheids that we have encountered earlier. These stars, known as *Miras*, or long-period variables, pulsate with periods of the order of a year.

As a consequence of the superwind, stars of initial mass in the range $1M_\odot < M < 9M_\odot$ shed their envelopes and are left with C-O cores of mass between $0.6M_\odot$ and $1.1M_\odot$, a higher final core mass corresponding to a higher initial total mass. These cores will subsequently develop into white dwarfs. Since, as we shall see shortly, low-mass stars are far more numerous than massive ones, we expect most white dwarfs to have masses near $0.6M_\odot$. This conclusion is verified by observations. Thus white dwarfs originating from AGB stars have masses considerably smaller than the Chandrasekhar critical mass, and hence, although degenerate, these stars are in no danger of a catastrophic denouement (contrary to some early theories). But they do undergo a short episode of particular brilliance before fading into cooling, inert white dwarfs.

The cores of stars at the end of the asymptotic giant phase are surrounded by an extended shell, a more or less spherical nebula formed by the ejected material. The inner part of this shell – resulting from the superwind – is relatively dense. When mass loss finally ceases, the core, freed from the burden of a massive envelope, expands slightly and as a result, the small envelope remnant contracts. This causes a distinct separation, a void, between the star and its ejecta. Subsequently, as the central star contracts, the effective temperature rises considerably. When it reaches $\sim 30\,000$ K, the radiated photons become energetic enough to ionize the atoms in the nebula and cause them to shine by fluorescence (the same mechanism that is responsible for fluorescent lamps). A shining nebula of this kind is called a *planetary nebula*; it appears as a bright circular ring surrounding a point-source of light, which, in the past, was thought to represent a disc of planets revolving around a central star, similar to the solar planetary system. An example is given in Figure 8.11. Obviously, if this were the case, at least some planetary nebulae should have appeared flattened, due to the inclination of the disc with respect to the line of sight to the observer. The fact that all planetary nebulae appear almost circular indicates that what we see is the projection of a spherical shell. As the line of sight through the nebula is much longer near the edges than at the centre, the material appears opaque toward the edge and transparent at the centre, making it possible to see the hot central star, as illustrated in Figure 8.12. This explains the ring shape. The central source is called the *planetary nebula nucleus*.

The path that planetary nebulae trace in the H–R diagram is a horseshoe-shaped track, first leftward, toward higher surface temperatures, meaning that the nucleus preserves its luminosity during the transition, and then downward and to the right. The energy is provided by nuclear burning in the thin shell still left on top of the C-O core. When the mass of this shell decreases below a critical size, of the order of 10^{-3} to $10^{-4} M_\odot$, the shell can no longer maintain the high temperature required for nuclear burning. The energy source becomes extinct, the luminosity of the central star drops, and its ionizing power diminishes. At the

8.7 The superwind and the planetary nebula phase

Figure 8.11 *Top*: the Helix nebula, the nearest (450 ly away) and largest observed planetary nebula [copyright Anglo-Australian Observatory; photograph by D. Malin]. *Bottom*: detail of the Helix nebula captured by NASA's Hubble Space Telescope, showing knots of gas. Each gaseous head is at least twice the size of our solar system and each tail stretches to about 1000 AU [photograph by C. R. O'Dell, Rice University].

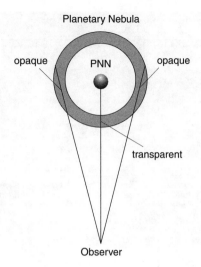

Figure 8.12 Sketch of a planetary nebula and its nucleus (PNN).

same time, the nebula, which expands at a rate of a few 10 km s^{-1}, grows in size and gradually disperses. Thus a planetary nebula fades away and disappears after some 10^4–10^5 yr. We now turn to the evolution of the remnant central star into a cool white dwarf.

8.8 White dwarfs – the final state of nonmassive stars

Most white dwarfs – compact stars of high surface temperature – descend from AGB stars, which develop carbon-oxygen electron-degenerate cores. As we have seen, these stars lose mass by a strong stellar wind, while undergoing thermal pulses caused by the alternate burning of hydrogen and helium in thin shells. The end of mass loss, brought about by the dissipation of the entire envelope, occurs at a random phase of a thermal pulse. If it occurs during the hydrogen burning phase, the star will be left with a thin coating of hydrogen-rich material, a vestige of the lost envelope. If it occurs during helium burning, which takes place at the bottom of a helium layer, the outer envelope will be composed predominantly of helium. Since helium burning takes up only a small fraction of the pulse cycle, the probability of a star ending the asymptotic giant stage with a helium, rather than a hydrogen-rich, envelope is proportionally smaller.

Soon after the end of mass loss, nuclear burning comes to an end as well. During the intervening, short-lived planetary nebula phase, the final stage of nuclear burning supplies the energy that lights up the ejecta of the former AGB star. The planetary nebula nucleus – the degenerate core of the former AGB star, with the

8.8 White dwarfs – the final state of nonmassive stars

remnant thin envelope – becomes a white dwarf. We should therefore expect to encounter two types of white dwarf spectra: a prevalent one, showing hydrogen lines, and a rarer type, with no evidence of hydrogen. Indeed, observations confirm this expectation: about 25% of white dwarfs have no hydrogen lines in their spectra.

Another source of white dwarfs are low-mass stars in a narrow initial mass range: $0.7 \lesssim M \lesssim 1 M_\odot$. These stars do not reach high enough temperatures so as to ignite helium, simply because they do not grow sufficiently massive helium cores. Following the main-sequence phase, they turn into red giants and lose most of their envelopes, while the cores grow – by shell hydrogen burning – to only $0.4 M_\odot$ or less. Skipping the core helium burning phase, the asymptotic giant phase and the planetary nebula phase, these low-mass stars become even lower-mass white dwarfs, composed mainly of helium. Indeed, the mass distribution of white dwarfs derived from observations shows two peaks. The main peak, to which most white dwarfs belong, corresponds to an average mass of $\sim 0.6 M_\odot$. The secondary, smaller, peak is found between $0.2 M_\odot$ and $0.4 M_\odot$, confirming the prediction of two distinct sources of white dwarfs.

How do white dwarfs evolve, as they must, since they radiate? This question was considered by Mestel in 1952. The structure of stars in the white dwarf stage is characterized by two basic properties:

1. The internal pressure is supplied predominantly by degenerate electrons.
2. The internal energy source responsible for the radiation emitted at the surface is the thermal energy stored by the ions (as the heat capacity of a degenerate electron gas is negligible). The star has no nuclear energy sources. If it had, nuclear burning would have been unstable (as seen in Section 6.2), and something would have happened to either stop it or disrupt the star.

Note: In fact, as a white dwarf cools, it does contract slightly, releasing some gravitational energy. At the same time, however, the higher density raises the internal energy of the degenerate electrons (for which $u \propto \rho^{2/3}$ – see Sections 3.3 and 3.5) and also the electrostatic potential energy. Mestel and Malvin Ruderman showed (in 1967) that, to first order, the release of gravitational energy compensates for the rise in degeneracy and electrostatic energy. Thus they vindicated the long-standing assumption that the energy source of white dwarfs is the thermal energy of the ions, as if the white dwarf were rigid.

A degenerate electron gas behaves much like a metal, conducting heat very efficiently. Since, by equation (5.3), a very low opacity value implies a very small temperature gradient, the internal temperature of a white dwarf is very nearly uniform. The white dwarf structure – a homogeneous, isothermal gas, with negligible

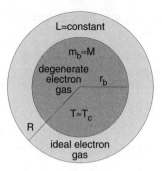

Figure 8.13 Sketch of the configuration of a cooling white dwarf.

radiation pressure and no nuclear reactions – appears simple enough to be described by analytical models with reasonable accuracy. (Elaborate numerical models are nevertheless required for supplying the finer structural details.)

A simple model for the *evolution* of a white dwarf is obtained following Mestel. A typical white dwarf may be described by an isothermal electron-degenerate core comprising most of the star's mass M. As the density decreases (tending to zero) toward the surface, an outer layer exists, however, where the electrons cease to be degenerate and behave as an ideal gas. Across this surface layer the temperature drops as well and radiative equilibrium may be assumed, with the temperature gradient determining the luminosity. The configuration bears similarity to that of a fully convective star, discussed in Section 8.1: there, too, the luminosity was determined by the conditions prevailing in a thin radiative outer layer (the photosphere). Obviously, the transition from a degenerate state to an ideal gas state is gradual, but, for simplicity, we shall assume a sharp transition across a surface boundary between the degenerate core and the ideal gas outer layer, defined as the point where the physical conditions are such that equal values result for the ideal gas pressure and for the degenerate gas pressure. Let the radius of this boundary be r_b, as shown in Figure 8.13. For $r < r_b$ the temperature is constant and equal to the central value T_c. For $r > r_b$ the luminosity is constant; in addition, $m(r > r_b) \approx M$. The structure equations for the outer layer reduce, therefore, to

$$\frac{dP}{dr} = -\rho \frac{GM}{r^2} \qquad (8.32)$$

$$\frac{dT}{dr} = -\frac{3}{4ac} \frac{\kappa \rho}{T^3} \frac{L}{4\pi r^2}. \qquad (8.33)$$

The first is derived from equation (5.1) with $m = M$, and the second is derived from equation (5.3) with $F = L$. For the opacity we shall assume a power-law

8.8 White dwarfs – the final state of nonmassive stars

dependence on temperature and density, the Kramers opacity law (3.65),

$$\kappa = \kappa_0 \rho T^{-7/2} = \frac{\kappa_0 \mu}{\mathcal{R}} P T^{-9/2}, \qquad (8.34)$$

where ρ has been replaced by P, using the ideal gas equation of state (3.28). Substituting equation (8.34) into equation (8.33) and dividing equation (8.32) by equation (8.33), we obtain a relation between the pressure and the temperature of the form

$$P \, dP = \frac{16\pi ac \mathcal{R} G}{3 \kappa_0 \mu} \frac{M}{L} T^{15/2} \, dT. \qquad (8.35)$$

Integrating from the surface, where $P = T = 0$, inward, we have

$$P(T) = \left(\frac{64\pi ac \mathcal{R} G}{51 \kappa_0 \mu} \right)^{1/2} \left(\frac{M}{L} \right)^{1/2} T^{17/4}. \qquad (8.36)$$

[This relation, which may be applied to the outer fringe (atmosphere) of stars in general, is known as the *radiative zero* solution.]

Reverting back to the density by means of the ideal gas equation of state, a relation is obtained between the density and the temperature

$$\rho(T) = \left(\frac{64\pi ac G \mu}{51 \mathcal{R} \kappa_0} \right)^{1/2} \left(\frac{M}{L} \right)^{1/2} T^{13/4}, \qquad (8.37)$$

which holds down to r_b. Since the ions constitute an ideal gas on both sides of r_b, it follows that r_b is the point where the ideal electron pressure, equation (3.27), and the degenerate electron pressure, equation (3.34), are the same. This leads to a second relation between density and temperature at r_b:

$$\left[\frac{\mathcal{R}}{\mu_e} \rho T \right]_b = \left[K_1' \left(\frac{\rho}{\mu_e} \right)^{5/3} \right]_b. \qquad (8.38)$$

Clearly, we must have $T_b = T_c$ in order to prevent a jump in temperature, which would result in an infinite heat flux. Eliminating ρ between equations (8.37) and (8.38), we finally obtain

$$\frac{L}{M} = \frac{64\pi ac G K_1'^3 \mu}{51 \mathcal{R}^4 \kappa_0 \mu_e^2} T_c^{7/2}, \qquad (8.39)$$

which relates the luminosity emitted at the surface to the core temperature of the white dwarf. Inserting the values of constants in equation (8.39) for a typical white dwarf composition (say, half carbon and half oxygen), we have

$$\frac{L/L_\odot}{M/M_\odot} \approx 6.8 \times 10^{-3} \left(\frac{T_c}{10^7\,\text{K}}\right)^{7/2} \tag{8.40}$$

or

$$T_c \approx 4 \times 10^7 \left(\frac{L/L_\odot}{M/M_\odot}\right)^{2/7} \text{K}. \tag{8.41}$$

> **Exercise 1.6:** (a) Show that the temperature profile throughout the outer layer of a white dwarf of mass M and radius R is given by
>
> $$T(r) = \frac{4}{17}\frac{\mu}{\mathcal{R}}GM\left(\frac{1}{r} - \frac{1}{R}\right). \tag{8.42}$$
>
> (b) Show that the layer's thickness $\ell \equiv R - r_b \ll R$. (c) Calculate the relative change in thickness, ℓ_1/ℓ_2, for a drop in luminosity from $L_1 = 10^{-2}L_\odot$ to $L_2 = 10^{-4}L_\odot$ (neglecting the small change in R).

As we have already mentioned, the energy source of a white dwarf is the thermal energy of the ions in the isothermal core (the outer layer's contribution being negligible):

$$U_I = \frac{3}{2}\frac{\mathcal{R}}{\mu_I}MT_c. \tag{8.43}$$

Hence the rate of energy emission L must equal the rate of thermal energy depletion:

$$L = -\frac{dU_I}{dt} = -\frac{3}{2}\frac{\mathcal{R}}{\mu_I}M\frac{dT_c}{dt} = -\frac{3}{7}\frac{\mathcal{R}}{\mu_I}M\frac{T_c}{L}\frac{dL}{dt}, \tag{8.44}$$

where we have used the $T_c(L)$ relation (8.39). It is easily shown that this implies

$$-\frac{dL}{dt} \propto MT_c^6, \tag{8.45}$$

which means that the rate of change of the luminosity (or, equivalently, the cooling rate) decreases sharply with decreasing temperature. Thus the evolutionary pace of a white dwarf slows down gradually, and a white dwarf of low mass evolves more slowly than a massive one. To estimate the time it would take a white dwarf of mass M to cool from an initial temperature T_c' (and corresponding luminosity

8.8 White dwarfs – the final state of nonmassive stars

L') to a temperature T_c (luminosity L), we integrate equation (8.44):

$$\tau_{\rm cool} = 0.6 \frac{\mathcal{R}}{\mu_{\rm I}} M \left(\frac{T_c}{L} - \frac{T_c'}{L'} \right). \tag{8.46}$$

If $T_c' \gg T_c$, then by equation (8.39), $T_c'/L' \ll T_c/L$ and the time required for a white dwarf to cool to a temperature T_c (from a much higher temperature) or decline to a luminosity L (from a much higher luminosity) is given by

$$\tau_{\rm cool} \approx 2.5 \times 10^6 \left(\frac{M/M_\odot}{L/L_\odot} \right)^{5/7} {\rm yr}. \tag{8.47}$$

For example, about 2×10^9 yr would be required for the luminosity of a $1 M_\odot$ white dwarf to drop to $10^{-4} L_\odot$. For comparison, only $\sim 10^7$ yr would bring the luminosity down to $0.1 L_\odot$ from, say, the typical planetary nebula luminosity, of the order of $10^4 L_\odot$.

In reality, when a white dwarf reaches very low temperatures (luminosities), the cooling rate no longer follows the simple relation (8.44). This is because the ion gas ceases to be perfect; Coulomb interactions increase in importance until, eventually, they become dominant. As the ratio ϵ_C/kT (discussed in Section 3.1) approaches and then surpasses unity, the ion gas crystallizes into a periodic lattice. At first, the corresponding heat capacity per ion increases due to the additional vibrational degrees of freedom (from $3/2k$ to $3k$). However, below a critical temperature (the *Debye temperature*), typically a few million degrees Kelvin, the heat capacity falls rapidly with temperature, following a T^3 law. This means that for a given amount of radiated energy, the drop in temperature is far larger than in the free gas regime. Thus the cooling of white dwarfs is accelerated considerably. If $\tau_{\rm cool} \propto L^\alpha$, then $\alpha = -5/7$ [equation (8.47)] holds down to $\sim 10^{-3} L_\odot$, with α increasing to small positive values below $\sim 10^{-4} L_\odot$. The number density of observed white dwarfs as function of their luminosity – shown in Figure 8.14 – bears witness to this effect.

The density distribution of a white dwarf, quite accurately described by an $n = 1.5$ polytrope for $M \lesssim 1.2 M_\odot$ (see Section 5.4), remains almost constant during the long cooling phase and hence so does the radius R. Therefore, the cooling track in the H–R diagram is essentially a $R = {\rm constant}$ (straight) line

$$\log L = 4 \log T_{\rm eff} + {\rm constant}, \tag{8.48}$$

the effective temperature decreasing with the luminosity (and almost linearly with the core temperature). Since $R = R(M)$, the evolution of white dwarfs of different masses corresponds to a strip in the H–R diagram, as shown in Figure 8.15. The lower part of this strip should be much more heavily populated than the upper part because of the rapidly decreasing cooling rate. White dwarfs spend far more

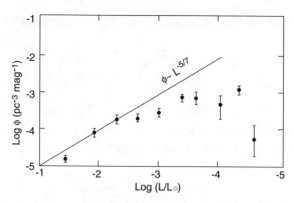

Figure 8.14 White dwarf luminosity function: number density of white dwarfs within a logarithmic luminosity interval corresponding to a factor of $10^{2/5} \approx 2.5$ against luminosity [data from D. E. Winget et al. (1987), *Astrophys. J.*, 315].

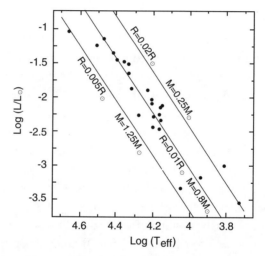

Figure 8.15 White dwarfs in the H–R diagram. Lines of constant radius (mass) are marked [data from M. A. Sweeney (1976), *Astron. & Astrophys.*, 49].

time at low luminosities than at high ones. These conclusions are confirmed by observations. Unfortunately, however, as they grow still fainter, they also become more difficult to detect (and besides, their number per luminosity interval drops due to the rapid cooling). In the end they will turn into practically invisible *black dwarfs*.

8.9 The evolution of massive stars

The evolution of massive stars ($M_0 > 10 M_\odot$) has the following general characteristics:

8.9 The evolution of massive stars

1. The electrons in their cores do not become degenerate until the final burning stages, when the core consists of iron.
2. Mass loss plays an important role along the entire course of evolution, including the main-sequence phase (as mass loss is still an elusive process, this is also the reason for the poorer understanding of these stars' evolution).
3. The luminosity, which is already close to the Eddington critical limit on the main sequence, remains almost constant, in spite of internal changes. The evolutionary track in the H–R diagram is therefore horizontal, shifting back and forth between low and high effective temperatures. Such transitions are slow during episodes of nuclear burning in the core and rapid during intervening phases, when the core contracts and heats up, while the envelope expands.

Stars of initial mass exceeding $30M_\odot$ have so powerful stellar winds as to result in mass loss timescales $M/\dot{M}$ shorter than main-sequence timescales MQ/L. Consequently, their main-sequence evolutionary paths converge toward that of a $30M_\odot$ star. In particular, the extent of the helium core at the end of the main-sequence phase is similar, and hence so are the ensuing evolutionary stages. The intense mass loss that occurs during the main-sequence phase leads to configurations composed mainly of helium, with hydrogen-poor envelopes ($X \approx 0.1$) or no hydrogen at all. Such stars – luminous, depleted of hydrogen, and losing mass at a high rate – are indeed observed, being known as *Wolf–Rayet stars*. They have relatively low average masses, between 5 and $10M_\odot$, and are considered as the bare cores of stars initially more massive than $30M_\odot$. There are different types of Wolf–Rayet stars, distinguished according to their surface composition. Element abundances in the sequence of types correspond to a progression in peeling off of the outer layers of evolving massive stars; thus, some show the undiluted burning products of the CNO cycle – helium and nitrogen, while others show the products of 3α and other helium burning reactions, mostly carbon and oxygen. A notorious example of vigorous mass loss is provided by the peculiar star η Carinae, shown in the upper panel of Figure 8.16. The nebula is considerably enriched in nitrogen, and generally the observed abundances are consistent with those obtained from model calculations for the supergiant phase of an initial $120M_\odot$ star evolving with mass loss. A recent image of mass ejection by a typical Wolf–Rayet star is shown in the lower panel of the figure.

In all massive stars, helium burning in the core is succeeded by carbon burning. At this stage the core temperature is so high as to cause significant energy losses that are due to neutrino emission. Thus the nuclear energy source has to compensate for these losses, as well as supply the high luminosity radiated at the surface. As fusion of heavy elements releases far less energy per unit mass of burnt material than fusion of light elements (see Chapter 4), nuclear fuels are very rapidly consumed. All the major burning stages pass in rapid succession, until an inner core made of iron group elements is formed. Surrounding this core are shells

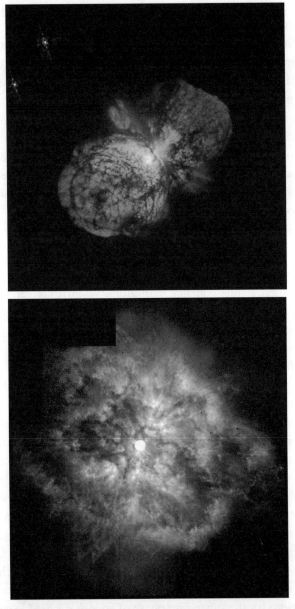

Figure 8.16 Mass ejection by massive stars captured by NASA's Hubble Space Telescope. *Top*: Eta Carinae, one of the brightest and most massive mass losing stars. Its luminosity is estimated at about $5 \times 10^6 L_\odot$, and its present mass at roughly $100 M_\odot$. Two lobes of ejected stellar material are located very near the star, moving outward at a velocity of ~ 600 km s^{-1} [photograph by J. Morse, University of Colorado]. *Bottom*: a massive, hot Wolf–Rayet star embedded in the nebula created by its intense wind. The blobs result from instabilities in the wind which make it clumpy. The expansion velocity is about 40 km s^{-1} and the nebula is estimated to be no older than 10^4 yr [photograph by Y. Grosdidier, University of Montreal and Observatoire de Strasbourg; A. Moffat, University of Montreal; G. Joncas, University of Laval; and A. Acker, Observatoire de Strasbourg].

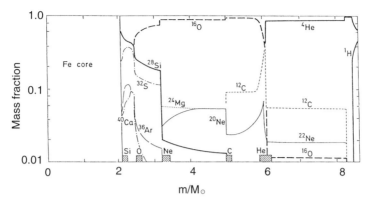

Figure 8.17 Composition profiles in the inner $8M_\odot$ of a $25M_\odot$ star prior to supernova collapse. Burning shells are marked [adapted from S. E. Woosley & T. A. Weaver (1986), *Ann. Rev. Astron. Astrophys.*, 24].

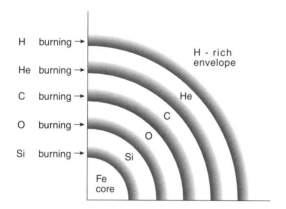

Figure 8.18 Schematic structure of a supernova progenitor star.

of different compositions – silicon, oxygen, neon, carbon, helium – and, finally, the envelope, which for $M_0 < 30 M_\odot$ retains most of the original composition and contains most of the stellar mass. The inevitable contraction of the iron core will lead the star toward collapse in a *supernova* explosion. The structure of a massive star and its schematic configuration in the supernova progenitor stage are shown in Figures 8.17 and 8.18. The final stages of evolution will be described in the next chapter.

8.10 The H–R diagram: Epilogue

We have come to the end of our discussion on the H–R diagram and its theoretical counterpart, the $(\log L, \log T_{\text{eff}})$ diagram, thus completing the task that we set out to accomplish at the end of Chapter 1. The success of the stellar evolution theory in explaining the many different, often puzzling, characteristics of stars, as exhibited

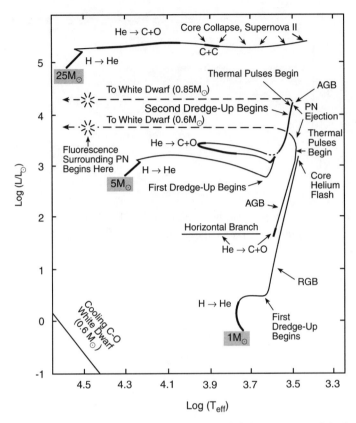

Figure 8.19 Evolutionary tracks of $1M_\odot$, $5M_\odot$, and $25M_\odot$ star models in the H–R diagram. Thick segments of the line denote long, nuclear burning, evolutionary phases. The turnoff points from the AGB are determined empirically [from I. Iben Jr. (1985), *Quart. J. Roy. Astron. Soc.*, 26].

by the H–R diagram, is remarkable: it explains the prevalence of main-sequence stars, the turning point off the main sequence in star clusters, the red giant, the supergiant, and the horizontal branches, the planetary nebula and the white dwarf regions, the gap between the main sequence and the giant branch, and many other, subtler properties of stars. To conclude this discussion, we show two more figures. In Figure 8.19 full evolutionary tracks in the H–R diagram are given for a low-mass star, an intermediate-mass star, and a massive star. Finally, crowning the stellar evolution theory, Figure 8.20 shows the evolving H–R diagram of a hypothetical star cluster, based on evolutionary calculations of a large number of star models of different masses performed by Rudolf Kippenhahn and Alfred Weigert. These are hardly distinguishable from actual H–R diagrams of stellar clusters of different ages, as shown in Chapter 1.

Nevertheless, the picture of stellar evolution is not yet complete, although it is far more elaborate, detailed, and clear than the rough sketch traced in the previous chapter. On close inspection, there are still fuzzy spots, especially where mass

8.10 The H–R diagram: Epilogue

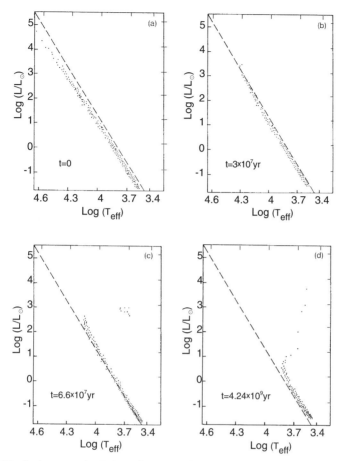

Figure 8.20 Evolutionary calculations for stars of different masses forming a hypothetical cluster result in an evolving H–R diagram, shown at four ages [adapted from R. Kippenhahn (1983), *100 Billion Suns*, Princeton University Press].

loss or convection are concerned. Eddington ended his famous 1926 book with the following: "... *but it is reasonable to hope that in a not too distant future we shall be competent to understand so simple a thing as a star.*"

We now do understand a great deal about stars; in particular, we understand that they are not all that simple.

9

Exotic stars: supernovae, pulsars, and black holes

Stars of the types considered in this chapter differ from those discussed so far, inasmuch as, for various reasons, they do not (or cannot) appear on the H–R diagram. As before, we shall rely on stellar evolution calculations to describe them. Whenever possible, we shall confront the results and predictions of the theory with observations, either directly or based on statistical considerations. We shall find that, as we approach the frontiers of modern astrophysics, theory and observation go more closely hand in hand.

9.1 What is a supernova?

We should start by making acquaintance with the astronomical concept of a *supernova*, as we did with main-sequence stars, red giants, and white dwarfs in Chapter 1. Stars undergoing a tremendous explosion (sudden brightening), during which their luminosity becomes comparable to that of an entire galaxy (some 10^{11} stars!), are called *supernovae*. Historically, *nova* was the name used for an apparently new star; eventually it turned out to be a misnomer, *novae* being (faint) stars that brighten suddenly by many orders of magnitude. So are *super*novae, but on a much larger scale. Not until the 1930s were supernovae recognized as a separate class of objects within novae in general. They were so called by Fritz Zwicky, after Edwin Hubble had estimated the distance to the Andromeda galaxy (with the aid of Cepheids) and had thus been able to appreciate the unequalled luminosity of the nova discovered in that galaxy in 1885, amounting to about one sixth of the luminosity of the galaxy itself.

Since supernova outbursts last for very short periods of time (several months to a few years), the chances of detecting them are small, even if an appreciable fraction of stars go through this stage. Thus in a large stellar population, such as a galaxy, supernova explosions are detected once in a few decades. Fortunately, they become bright enough to be observable at very large, cosmic distances, and hence hundreds of such events have been recorded and studied. An example is given in

9.1 What is a supernova?

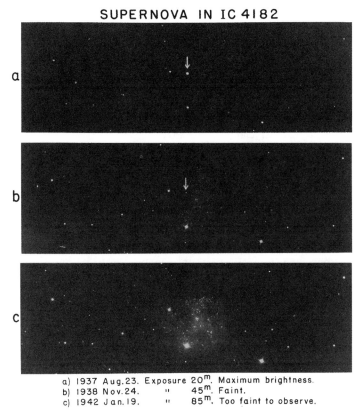

a) 1937 Aug.23. Exposure 20^m. Maximum brightness.
b) 1938 Nov.24. " 45^m. Faint.
c) 1942 Jan.19. " 85^m. Too faint to observe.

Figure 9.1 Supernova in the galaxy IC4182. At maximum brightness (*top*), it completely obscures the galaxy; 5 years later (*bottom*) it becomes too faint to observe and the parent galaxy appears in the picture [Mt. Wilson 100-in. telescope photographs from the Hale Observatories].

Figure 9.1, where a supernova in outburst outshines the galaxy within which it resides by about a factor of 100. The most notorious supernovae are those which occurred and were observed in our own Galaxy – the historical supernovae, listed in Table 9.1. These, however, represent only a fraction of *all* supernova explosions that must have occurred in our Galaxy, say, in the last millennium, because most regions of our Galaxy are obscured by its radiation-absorbing central bulge. (It is much easier to detect lights turning on in a neighbouring building than in one's own.)

Historical Note: The close occurrence of the supernovae of 1572 and 1604 led to a philosophical revolution, by shattering the Aristotelian conception of the universe, which had prevailed for almost two millennia. Aristotle's universe consisted of a set of concentric spheres, with the Earth at the centre. Each of the planets known at the time revolved in its own sphere, while the outermost sphere contained the fixed stars. The lowest sphere contained the Moon and marked

Table 9.1 Historical supernovae

Galaxy: Name	Year	Distance (3000 ly)
Milky Way:		
Lupus	1006	1.4
Crab	1054	2.4
3C 58	1181(?)	2.6
Tycho	1572	2.5
Kepler	1604	4.2
Cas A	1658 ± 3	2.8
Andromeda	1885	700
LMC: SN1987A	1987	50

the boundary between the imperfect, changeable world below it and the perfect and eternal universe above. This is the reason why comets, of unpredictable and transient apparition, were considered atmospheric phenomena. The supernova of 1572 was intensively observed and studied by the Danish astronomer Tycho Brahe, who devoted a book (*De Nova Stella*) to the new star. He paid particular attention to its distance and concluded that it must reside within the fixed stars, far above the Moon, showing that changes could take place in what had been considered the immutable universe. But he chose to explain the new star as an immutable object that had so far been concealed from the human eye. It took one more (soon to follow) supernova, another great astronomer – Johannes Kepler, Tycho's former assistant – and one more publication (bearing a similar title, *De Stella Nova*) to overthrow the conception of the immutability of the heavens. Kepler observed the 1604 supernova and concluded that, like Tycho's supernova, it, too, was among the fixed stars. Aristotle's model had failed again and was soon to be abandoned, although reluctantly at first, in favour of the Copernican heliocentric theory and Kepler's famous laws of planetary motion.

The nebulae ejected in supernova explosions, the so-called *supernova remnants*, survive for much longer periods of time and are regarded as some of the most spectacular astronomical objects. Expansion velocities being very high, up to 10 000 km s^{-1} (0.03c!), these nebulae extend quite rapidly to remarkable dimensions and remain visible for thousands of years, even when they become so dilute as to be almost transparent. The Crab nebula, a remnant of the 1054 supernova, is shown in Figure 9.2. A much more diluted, older supernova remnant is shown in Figure 9.3.

According to the stellar evolution *theory* that we have outlined in Chapter 7, a catastrophic end to the life of a star is bound to arise from two types of quite different circumstances, which lead to a state of dynamical instability. One is the collapse of the iron cores of massive stars. The other is the collapse of white dwarfs that have reached the Chandrasekhar limiting mass. As we have seen in the previous chapter, the masses of degenerate cores of intermediate-mass stars,

9.1 What is a supernova?

Figure 9.2 Crab nebula: the expanding remnant of the supernova that exploded in 1054 [from plates taken in 1956 with the Hale 5m telescope, copyright D. Malin and J. Pasachoff, Caltech].

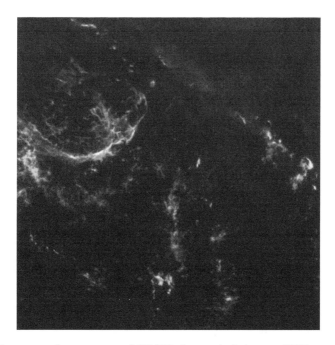

Figure 9.3 Remnant of a supernova (N132D) that exploded some 3000 years ago in the Large Magellanic Cloud. The progenitor star, which was located slightly below and left of centre in the image, is estimated to have had a mass of $25 M_\odot$ [photograph by J. A. Morse, Space Telescope Science Institute, taken with NASA's Hubble Space Telescope].

which turn eventually into white dwarfs, are considerably lower than the critical mass. Hence single stars are spared the catastrophic fate of collapse. This fate awaits, however, white dwarfs evolving in binary systems, which accrete material from a companion star and may reach M_{Ch}.

Indeed, supernova explosions are classified into two types according to their *observed* properties: the so-called *Type I* and *Type II* supernovae. The main distinguishing characteristic is the presence of hydrogen lines in the spectrum of the latter and their absence in the former. Each type has its own characteristic light curve, although a wide variety of deviations from the general shape is detected, resulting from individual properties, and subclasses have been defined (which we shall ignore). Type II supernovae are not observed in old stellar populations (such as elliptical galaxies), but mostly in the gas and dust rich arms of spiral galaxies, where star formation is going on and young stars are abundant. Type I supernovae, by contrast, are found in all types of galaxies.

It is the Type II supernovae that are associated with the collapse of the iron cores of massive stars. These stars have large hydrogen-rich envelopes; hence the evidence of hydrogen in the spectrum. As massive stars evolve much more rapidly than low-mass stars, old stellar populations, where no star formation occurs, have outgrown the Type II supernova stage. Type I supernovae are those believed to arise from the collapse of white dwarfs that have reached the Chandrasekhar limiting mass, presumably by accretion. Since in a given stellar population white dwarfs form at all times, and since accretion rates may widely vary, there is nothing to prevent the occurrence of Type I supernovae in old populations as in young. As we set out to explore the evolution of only single (isolated) stars, we shall presently focus on the final stages of evolution of massive stars, the Type II supernova precursors.

9.2 Supernova explosions – the fate of massive stars

To summarize Section 8.9, stars of initial mass exceeding $\sim 10 M_\odot$ undergo all the major burning stages, ending with a growing iron core surrounded by layers of different compositions. These are separated by burning fronts, which turn the lighter nuclear species of the overlying layer into the heavier species of the underlying one. Anticipating the imminent collapse, we have called such stars *supernova progenitors*.

At the beginning, the iron core contracts – as all inert stellar cores do – simply because no nuclear burning is taking place and, eventually, the electrons become a degenerate gas. When the degenerate core's mass surpasses the Chandrasekhar limit (which, for iron, is somewhat lower than $1.46 M_\odot$), the degenerate electron pressure is incapable of opposing self-gravity and the core goes on contracting rapidly. Two types of instability soon develop. First, electron capture by the heavy nuclei deprives the core of its main pressure source and thus accelerates the infall. Secondly, due to the high degeneracy of the gas – and hence its low sensitivity to

9.2 Supernova explosions – the fate of massive stars

temperature – the temperature rises unrestrained. In time, it becomes sufficiently high for the photodisintegration of iron nuclei (see Section 4.10):

$$^{56}\text{Fe} \longrightarrow 13\,^{4}\text{He} + 4\text{n} - 100 \text{ MeV}.$$

This reaction is highly endothermic, absorbing ~ 2 MeV per nucleon (just as the reverse transition of helium into iron releases ~ 2 MeV per nucleon). The loss of energy is so severe as to turn the collapse into an almost free fall. The continued contraction is followed by a further rise in temperature. The pressure increases too, but not sufficiently for arresting the process ($\gamma_a < 4/3$). The infall continues until the photons become energetic enough to break the helium nuclei into protons and neutrons. As this reaction entails an even greater energy absorption, about 6 MeV per nucleon, the core contracts still further. Eventually, the density becomes high enough for the free protons to capture the free electrons and turn into neutrons. Not only does this process absorb energy, but it also reduces the number of particles. Hence the pressure drops and core collapse continues. Finally, the neutron gas, which is in many ways similar to an electron gas, becomes degenerate. This occurs at a density of about 10^{18} kg m^{-3} (10^{15} g cm^{-3}) and generates sufficient pressure to halt the collapse. A neutron core is thus created, of a density similar to that of an atomic nucleus – one single huge nucleus, about 40 km in diameter. It was Hoyle who, as early as 1946, suggested the instability associated with the photodisintegration of iron to be the triggering mechanism for supernova explosions.

Exercise 9.1: Assuming a (free-fall) collapsing core to maintain a uniform density (this is called *homologous contraction*), show that the solution of the equation of motion tends to $|v| \propto r$.

Exercise 9.2: Show that the free-fall collapse of a stellar core (of uniform *initial* density) is homologous.

What happens to the outer layers of the star during and following the few hundred milliseconds of core collapse? To answer this question, we consider the energy budget of the star. Clearly, the energy source of a supernova explosion is gravitational: the collapse of a core of mass $M_c \sim 1.5 M_\odot$ from an initial white dwarf radius R_c ($\sim 0.01 R_\odot$) to the final radius $R_{nc} \sim 20$ km ($\ll R_c$) of the neutron core releases an amount of gravitational energy of the order of

$$\Delta E_{\text{grav}} \approx -GM_c^2 \left(\frac{1}{R_c} - \frac{1}{R_{nc}} \right) \approx \frac{GM_c^2}{R_{nc}} \approx 3 \times 10^{46} \text{ J}. \tag{9.1}$$

The energy absorbed in nuclear processes amounts to

$$\Delta E_{\text{nuc}} \approx 7 \text{ Mev} \frac{M_c}{m_H} \approx 2 \times 10^{45} \text{ J}, \tag{9.2}$$

about one tenth of ΔE_{grav}. There remains ample energy for ejecting all the material outside the core, for imparting to it enormous velocities, and for producing the huge luminosities observed. The radiated energy may be estimated by assuming a typical luminosity $L_{\text{SN}} \sim 10^{37}$ J s^{-1} ($\sim 3 \times 10^{10} L_\odot$) for a typical period τ_{SN} of one year:

$$\Delta E_{\text{rad}} \approx L_{\text{SN}} \tau_{\text{SN}} \approx 3 \times 10^{44} \text{ J}. \tag{9.3}$$

Although this may be an overestimate, it is still only a few percent of the released energy. A similar amount would be required for the ejection of the (for the most part) loosely bound envelope,

$$\Delta E_{\text{bind}} \approx \frac{GM_c(M - M_c)}{R_c} \approx 5 \times 10^{44} \text{ J}, \tag{9.4}$$

assuming a total stellar mass $M \sim 10 M_\odot$, and a comparable amount would suffice for supplying the high expansion velocities of the ejecta:

$$\Delta E_{\text{kin}} \approx \tfrac{1}{2}(M - M_c)v_{\text{exp}}^2 \approx 10^{45} \text{ J}, \tag{9.5}$$

adopting $v_{\text{exp}} \sim 10\,000$ km s^{-1}, as derived from observations.

Two questions immediately arise: first, if such a small fraction of the released energy is sufficient for powering a supernova explosion, where does the bulk of the energy go? Second – the question that has puzzled astrophysicists for decades – what is the mechanism that deposits the required energy in the envelope? The answers to these questions are linked and involve one of the major factors affecting the entire supernova process that we have yet to mention. These are the neutrinos, which take part in any weak interaction so that the lepton number be conserved (see Section 2.6).

As the iron core turns, essentially, into a neutron core, all the protons that have been locked up in the iron nuclei undergo a weak interaction. Hence as many as 10^{57} neutrinos are released, which can easily remove $\sim 10^{46}$ J of energy. The second question is then merely how to transfer a small fraction of the neutrino energy to the envelope surrounding the collapsing core. Keeping in mind that matter is normally highly transparent to neutrinos, this has proved to be a very intricate question. Given, however, the enormous neutrino flux and the unusually high densities involved, it turns out that a nonnegligible neutrino opacity builds up. Some of the neutrino energy is absorbed by the envelope layers that bounce off the stiffened neutron core and are thus precipitated outward. The release of gravitational energy as the primary energy source in supernova explosions as well as the transfer of energy to the mantle by neutrinos were first proposed and studied by Stirling Colgate and Richard White in 1966. Recent numerical simulations – which include extensive, often multi-dimensional calculations performed on the

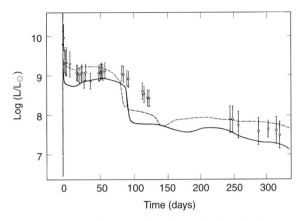

Figure 9.4 Light curves resulting from calculated models of a $15 M_\odot$ supernova compared with observations of SN1969I. The models differ in magnitude of the explosion energy: 1.3×10^{51} erg (solid line) and 3.3×10^{51} erg (dashed line) [adapted from A. J. Weaver & S. E. Woosley (1980), *Ann. NY Acad. Sci.*, 336].

most efficient computers – are quite successfully accounting for the observed characteristics of supernova explosions.

The flare-up of the supernova begins when the shock wave propagating from the collapsed core boundary breaks out through the surface of the hydrogen-rich envelope. At first, the temperature is so high that most of the energy is radiated in the UV, but very soon the envelope expands and the temperature drops sufficiently for the object to become visible. A typical light curve of a Type II supernova, where calculated models are superimposed on observational data points, is shown in Figure 9.4.

A unique opportunity to test the core-collapse-neutrino-generating theory was provided by the supernova that exploded in February of 1987 (known as SN1987A) in the *Large Magellanic Cloud* (LMC), a nearby galaxy, about 170 000 light-years away. A few of the neutrinos produced by that supernova (170 000 years ago) – to be exact, 20 out of an estimated 10^{13} per m^2 – were intercepted by the neutrino detecting devices Kamiokande (described in Section 8.3) and a similar one, named IMB, located in a 1570-m deep salt mine in Ohio. The first detections of the two widely separated devices were simultaneous to within the accuracy of time determination, and the entire neutrino capture event lasted about 12 s. It is noteworthy that since the detectors are located in the Northern Hemisphere, the neutrinos from the LMC traversed the Earth before hitting the detectors from below. All this occurred several hours *before* the supernova became visible, as the theory would have it, for some time must elapse following the collapse until the envelope expands enough to produce the typical supernova luminosity. Besides providing the first detected neutrinos associated with core collapse, SN1987A was unique in another, quite different sense: it was the first (and so far *only*) supernova whose progenitor had been identified and its location in the H–R diagram established

($\log T_{\text{eff}} = 4.11 - 4.20$, $\log L = 5.04$). A mass of $\sim 18 M_\odot$ was inferred (having probably evolved from a star of initial mass somewhat above $20 M_\odot$), in good agreement with the outburst and postoutburst characteristics. The supernova near maximum brightness is shown in Figure 9.5; superimposed is the negative of the progenitor star. All other supernovae we have known were either too distant or too old for their progenitors to have been distinguishable.

9.3 Nucleosynthesis during supernova explosions

Perhaps the most important and long-lasting outcome of supernova explosions is the production of heavy elements (heavier than helium) and their dispersion throughout the interstellar medium. These elements are produced both during the stages preceding the explosion, in the layers surrounding the iron core, and during the explosion itself, as a result of the shock wave that sweeps the mantle. Most of the shock wave energy turns into heat, which raises the temperature to peak values attaining 5×10^9 K; at such high temperatures nuclear statistical equilibrium is achieved (see Section 4.7) on a timescale of seconds (the dynamical timescale). The main product is ^{56}Ni, rather than iron, which is obtained at lower temperatures, when nuclear reactions are slower. The reason is that the nuclear fuel has $\mathcal{Z}/\mathcal{A} \approx \frac{1}{2}$, and since time is too short for β decays to occur and change the ratio of protons to neutrons, the product must also have $\mathcal{Z}/\mathcal{A} = \frac{1}{2}$, as ^{56}Ni does, whereas for ^{56}Fe, $\mathcal{Z}/\mathcal{A} = \frac{26}{56} < \frac{1}{2}$. As the shock wave moves out, it loses energy and its temperature declines. When the temperature falls below $\sim 2 \times 10^9$ K, which occurs when the wave has reached the neon-oxygen layer, explosive nucleosynthesis ceases. Thus elements heavier than magnesium are produced during the supernova explosion, while lighter elements are produced during the stages preceding it.

Typical values for the estimated ejected mass, as well as other characteristic masses of supernova models, are given in Table 9.2. The supernova ejecta mix with the pre-existing interstellar clouds made predominantly of primordial hydrogen and helium and thus determine the evolving galactic (cosmic) abundances of the elements. We shall return to this point in the next chapter. We only note for now that the agreement between the calculated ejecta abundance pattern and the solar system abundance pattern is striking, all the more so when one considers the span of seven orders of magnitude among the different species.

The production of ^{56}Ni, which is radioactive with a half-life of 6.1 days, has a marked effect on the supernova light curve and can therefore be verified by observations. The product of ^{56}Ni decay is ^{56}Co, itself radioactive with a half-life of 77.1 days, decaying into ^{56}Fe. These β decays release the energy (3.0×10^{12} J kg^{-1} for ^{56}Ni and 6.4×10^{12} J kg^{-1} for ^{56}Co) that powers the supernova light curve after the initial decline from maximum. As the rate of decay and energy release

9.3 Nucleosynthesis during supernova explosions

Figure 9.5 *Top*: SN1987A in the LMC, before and after outburst. *Bottom*: SN1987A in the LMC photographed in March 1987, about a month after discovery. Overlaid on the picture is the negative image taken a few years before. The image of the supernova progenitor is confused with two other stars in the same line of sight and thus appears noncircular [copyright Anglo-Australian Observatory; photographs by D. Malin].

Table 9.2 Characteristic masses of supernova models (in $M_\odot$)

Initial Mass	Helium Core	Iron Core	Neutron Core	Ejected ($Z \geq 6$)
15	4.2	1.33	1.31	1.24
25	8.5	2.05	1.96	4.31

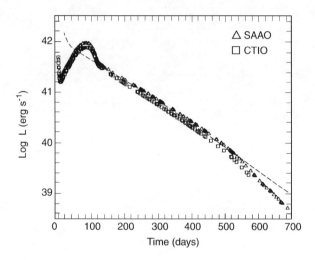

Figure 9.6 Light curve of SN1987A. Points correspond to observational data obtained at the Cerro Tololo Inter-American Observatory (CTIO) and the South African Astronomical Observatory (SAAO). The dashed line is obtained from a model assuming decay of $0.075 M_\odot$ of ^{56}Ni and later ^{56}Co [from D. Arnett et al. (1989), *Ann. Rev. Astron. Astrophys.*, 27].

decline exponentially on the appropriate timescales, it can be compared with the rate of decline of the light curve. A perfect match is obtained, as shown in Figure 9.6 for SN1987A. If the distance to the supernova is known, as it is in the case of SN1987A, the amount of ^{56}Ni produced can be inferred ($0.075 M_\odot$ for SN1987A). This effect is even more conspicuous in the light curves of Type I supernovae, where a much larger fraction of the mass, almost the entire progenitor star, turns into ^{56}Ni by explosive nucleosynthesis. The decay of ^{56}Ni and ^{56}Co dominates the light curve in that case. A typical Type I supernova light curve is shown in Figure 9.7.

A rather recent observation that bears direct witness to ongoing nucleosynthesis and to the continual dispersion of nuclear ashes throughout the Galaxy is the detection of ^{26}Al – a radioactive isotope with a half-life of 7.2×10^5 yr – in the interstellar medium. Nuclei, like atoms, have their distinct spectra, due to their discrete energy levels, only the energies involved are more than three orders of magnitude higher. Excited nuclei, like excited atoms, emit photons whose energies, corresponding to the differences between nuclear energy levels, are measured in

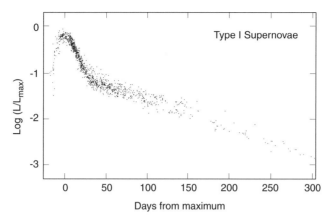

Figure 9.7 Composite light curve of 38 Type I supernovae (superimposed so that the maxima coincide) [from R. Barbon et al. (1973), *Astron. Astrophys.*, 25].

MeV, rather than eV (or keV), as in the case of atoms. The radioactive aluminium isotope is produced by nuclear reactions at high temperatures in an excited state. Its subsequent decay to ^{26}Mg releases a characteristic 1.8-MeV photon, detected as a line in the γ-ray spectrum. The detection of this line – by satellites equipped with the appropriate instruments – indicates that ^{26}Al has been produced in quite significant amounts no earlier than 10^6 yr ago, a very short time on the stellar evolution scale. Moreover, traces of the daughter product have been detected in meteorites. This means that the same process of nucleosynthesis and heavy element dispersion was taking place at the time when the solar system formed, 4.6×10^9 yr ago.

9.4 Supernova progenies: neutron stars – pulsars

With the expulsion of the envelope in a supernova explosion, the neutron core becomes a neutron star. The existence of such exotic objects as neutron stars was first postulated by Lev Landau, as early as 1932 (more precisely, Landau mentioned the possible formation of "one gigantic nucleus," when atomic nuclei come in close contact in stars exceeding the critical mass). Their resulting from supernovae was soon suggested by Walter Baade and Fritz Zwicky, in 1934, and the first physical model was offered by Robert Oppenheimer and George Volkoff in 1939. The governing equation of state is similar to that appropriate to a degenerate electron gas, a $n = 1.5$ polytrope (see Section 5.4) leading to a relation between mass and radius $R \propto M^{-1/3}$, so long as relativistic effects are negligible. For example, a $1.5 M_\odot$ neutron star would have a 15-km radius. Thus, whereas a white dwarf is similar in size to the Earth, the diameter of a neutron star is no bigger than that of a large city. As in the case of degenerate electrons, the relativistic limit of the equation of state imposes an upper limit on the neutron star's mass (equivalent to the Chandrasekhar limiting mass for white dwarfs derived in Section 5.4). Above this critical mass, a neutron star would not be able to generate enough pressure for balancing self-gravity, and collapse would ensue. In the case of neutrons, however, this limiting mass is far more difficult to estimate. The value of $5.83 M_\odot$ that would result from equation (5.32), by taking $\mu_n = 1$ for neutrons, rather than $\mu_e = 2$ for electrons, is incorrect for two reasons. First, in a relativistic neutron gas the kinetic energy of the particles is comparable with the rest-mass energy, and hence the Newtonian gravitational theory is no longer valid and Einstein's General Theory of Relativity (1915) must be used instead. Secondly, the gas is imperfect and particles can no longer be considered free (noninteracting) at the high neutron star densities. Interparticle distances are of the order of the strong force range. Hence nuclear forces have to be taken into account and the equation of state becomes more difficult to calculate. Although the first correction lowers the upper limit to about $0.7 M_\odot$, the second correction raises it. Thus, depending on the equation of state used, the upper limit to the neutron star mass is estimated to lie between $2 M_\odot$ and $3 M_\odot$. Fortunately, this limit does not impose serious constraints, since the iron cores of massive stars do not appear to exceed $2 M_\odot$ by much (see Table 9.2). And yet, in principle at least, a third end state would be possible for extremely massive stars – the collapse of a too massive (neutron) core into a *black hole*.

In all cases considered so far, the different types of astronomical objects had been observed long before they were understood. Such are main-sequence stars, red giants, white dwarfs, planetary nebulae, different types of variable stars, novae, supernovae – many misnomers bearing witness to that. Neutron stars, on the other hand, first emerged as theoretical, hypothetical objects. Then, in 1967, an important discovery was made, quite accidentally. Jocelyn Bell, a doctoral student working under the supervision of Anthony Hewish at a new radio telescope in Cambridge, detected variable radio sources of extremely high and very regular frequencies, the first such source having a period of barely more than 1 s. They were called *pulsars*, short for *puls*ating st*ars*, and by analyzing the pulses more closely, it was very soon realized that pulsars must be very compact galactic objects, smaller and denser than white dwarfs. To stress the impact of this discovery, we note that the 1974 Nobel Prize for Physics was awarded to Hewish for "*his decisive role in the discovery of pulsars*" (the prize was shared with Martin Ryle, for pioneering work in radio astronomy). Although the explanation for these strange objects was soon to follow, it was not soon enough to prevent another misnomer. Pulsars are *not* pulsating stars. As far as we can tell, they are *rotating neutron stars* – as suggested by Thomas Gold in 1968 – that formed in supernova explosions.

The association of pulsars with supernovae – suggested by Hoyle as soon as pulsars were made known – became widely accepted with the discovery of the notorious *Crab Pulsar*, with a period of only 0.033 s, at the centre of the nebula bearing the same name, which had already been identified as the remnant of the 1054 supernova. Another supernova remnant pulsar is the *Vela Pulsar*, with a

9.4 Supernova progenies: neutron stars – pulsars

frequency of 0.089 s, discovered soon afterward within the dispersed nebula of a supernova that occurred some 10 000 years ago.

Having identified pulsars as the neutron star survivors of supernova explosions, we now turn to the pulsar *mechanism*. This invokes two main factors: rotation and a magnetic field, both having been tremendously intensified in the course of the supernova core collapse. Taking into account that all stars rotate, even if very slowly and insignificantly, we may estimate the expected rotation rate of a neutron star by applying the law of angular momentum conservation to the collapsing core. Since the collapsing core is weakly coupled to the mantle, the transfer of angular momentum between them (if any) is negligible. We know that the Sun rotates with a period of 27 days. Let us assume this period $\mathcal{P}_\odot$ to be typical of a $1 R_\odot$ object of mass similar to the Sun's, such as the pre-supernova core of mass M_c. The corresponding angular momentum is of the order of

$$\mathcal{L} \approx \frac{2\pi M_c R_\odot^2}{\mathcal{P}_\odot}. \tag{9.6}$$

If the angular momentum is conserved while the core turns into a neutron star of radius $R_{ns} \simeq 20$ km, the rotation period of the neutron star will be

$$\mathcal{P}_{ns} = \mathcal{P}_\odot \left(\frac{R_{ns}}{R_\odot} \right)^2 \approx 2 \times 10^{-3} \text{ s}, \tag{9.7}$$

of the same order as the pulse period of the fastest pulsars.

But rotation alone is not sufficient for sending out pulsed radiation; a *beam* would be required that would be periodically directed toward the observer, similarly to a lighthouse beam. Here the magnetic field, which has been enhanced by the collapse up to 10^8 T – nine orders of magnitude higher than the Sun's and more than four orders of magnitude higher than a white dwarf's – comes into play. Charged particles are accelerated by a magnetic field, and accelerated charged particles emit radiation. Radiation of this kind, characteristic of strong magnetic fields, such as those produced in particle accelerators called *synchrotrons*, is duly called *synchrotron radiation*. It is mainly produced by electrons and it is strongest along the direction of the magnetic poles, where the magnetic field lines are most concentrated. Thus the radiation emitted by the neutron star comes from two radiating cones around the poles of the magnetic (dipole) axis, as shown in Figure 9.8. If the rotation axis and the magnetic field axis were to coincide, the neutron star, visible to observers lying within the radiation cones, would have appeared as a constant source. There is no reason, however, for the axes to be aligned – on Earth, for example, they are not. If they are not, we have our rotating beam that sweeps the sky at the frequency of the stellar spin. It should be noted that, since pulsars radiate in a preferred direction, we miss many such objects, even if they are close by.

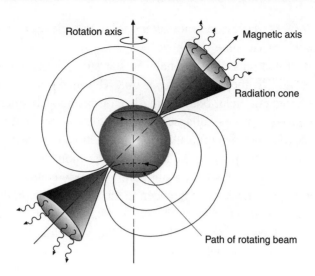

Figure 9.8 Sketch of the lighthouse model for pulsars.

What is the energy source of pulsars? The perhaps surprising answer is *kinetic energy* – the kinetic energy of rotation, amounting initially to

$$\Delta E_{\text{rot}} \approx \frac{1}{2} M_c \frac{4\pi^2 R_{\text{ns}}^2}{\mathcal{P}^2} \approx 5 \times 10^{45} \text{ J}, \qquad (9.8)$$

to use the values of the former example. [This energy, whose source is the gravitational energy of collapse, should have been added, in fact, to the energy budget of the supernova – equation (9.5) above – but would not have changed any of the conclusions.] Indeed, now that over a thousand pulsars have been detected in our Galaxy, and the oldest have been observed for a relatively long period of time, it has been established that the pulsars' periods increase with time (as Gold had predicted), implying that the spinning rate slows down. The rate of rotational energy loss $\dot{E}_{\text{rot}}$ derived from the observed slowdown of the spinning rate, $-\dot{E}_{\text{rot}} \propto \dot{\mathcal{P}}/\mathcal{P}^3$, exceeds by many orders of magnitude the rate of emission of pulsed radiation (a factor of $\sim 10^7$ for the Crab pulsar), so most of the energy is emitted by a different mechanism. In simple terms, the rapidly changing magnetic field of a spinning magnetic dipole with unaligned spin and dipole axes generates a strong electric field and emits electromagnetic radiation at the spin frequency, known as *magnetic dipole radiation*.

The magnetic dipole radiation mechanism for pulsars helps solving a long-standing enigma related to the very source of energy that powers the Crab nebula and, in particular, to that part of the radiation emitted by the nebula which is due to relativistic electrons. Although such electrons were produced in the supernova explosion, they should have radiated away their energy a long time ago. Moreover, an initial magnetic field that might have permeated the supernova ejecta, and could have accelerated these electrons, should have weakened considerably with the expansion of the nebula. As it turns out, for the Crab pulsar, $-\dot{E}_{\rm rot} \sim 10^5 L_\odot$, very close to the power required to explain the radiation and expansion of the Crab nebula. If the rotation slowdown is due to emission of magnetic dipole radiation, the relativistic electrons are a by-product of the huge electric field associated with the rapidly changing magnetic field of the spinning pulsar. The pulsar radiation and the transfer of energy from the pulsar to the nebula are not yet well understood, but, quite remarkably, John Archibald Wheeler and Franco Pacini had suggested that the Crab nebula might be powered by the magnetic dipole radiation of a rotating neutron star a short time *before* pulsars were discovered.

From the known number of pulsars and their estimated lifetimes it is possible to derive an average rate of pulsar formation: this turns out to be about one every few decades, very close to the rate of supernova explosions. This observation provides an indirect, but independent, corroboration for the association of pulsars with supernovae.

Finally, as their energy source wears out (after some 10^5–10^6 yr), pulsars, too, are destined to go into oblivion.

9.5 Very massive stars and black holes

Stars of mass $M \gtrsim 60 M_\odot$ (or $M \gtrsim 80 M_\odot$, according to some estimates) encounter a different type of instability in their evolutionary course. The brief hydrogen burning phase is followed, as usual, by helium burning. Helium burning in these stars produces mostly oxygen (due to the high core temperatures attained) and hence oxygen rather than carbon constitutes the next nuclear fuel. Oxygen ignites in a core exceeding $30 M_\odot$ at a temperature of $\sim 2 \times 10^9$ K. At this temperature the photon energy is sufficiently high for spontaneous electron-positron pair creation (see Section 4.9). Pair production, much as ionization or photodisintegration, reduces the adiabatic exponent below the stability limit of 4/3, leading to a dynamical instability, as discussed in Section 6.4. The core (or part of it) – whose mass exceeds the limiting neutron star mass – collapses and a black hole is formed on a dynamical timescale.

The description of a black hole, in fact the very concept of such an object, is entirely based on the theory of general relativity, which is beyond the scope of this text. Suffice it to say that even simple arguments indicate that something odd must occur when the radius of a star of given mass becomes so small that the escape velocity approaches the speed of light. This limiting radius, known as the *Schwarzschild radius* (after Karl Schwarzschild), is given by

$$R_{\rm Sch} = \frac{2GM}{c^2} \simeq 3 \frac{M}{M_\odot} \text{ km.} \quad (9.9)$$

(With remarkable foresight, Laplace pointed out the difficulty in 1795, about a century after Newton's classical theory of gravitation, but also about a century before Einstein's relativistic one.) Classical mechanics cannot account for such an enormous gravitational field as obtains at $r \leq R_{\text{Sch}}$ to prevent photons from escaping it, since photons are massless, but lacking more advanced knowledge, this is the intuitive explanation for the *blackness* of a black hole.

While the core collapses, the outer layers of the star are ejected as in a supernova explosion. Whether the explosion resembles a Type I or a Type II supernova or whether it is different from both is uncertain. The results of evolutionary calculations depend on as yet undetermined parameters, such as the mass loss rate during early evolution. It is unclear, for example, whether such a star would lose all, or only part of, its hydrogen-rich envelope prior to explosion. Nor can we rely on observations to guide us, for, as we shall see shortly, very massive stars are rarely born, and moreover, they are extremely short-lived, aging and dying, as it were, almost as soon as they are born. The remnant black hole, as the name indicates, would be elusive to observation as well. We have come to an impasse: we can never be certain of an explanation to a phenomenon that we do not see, even if we *can* explain *why* we do not see it. A theory which cannot be tested or confirmed by observation seems unworthy of pursuit. Nevertheless, in some cases there is compelling evidence for the presence of black holes (of stellar mass), inferred from phenomena related to the strong gravitational fields these objects generate around them. As such phenomena necessarily involve stellar interactions, they are beyond the scope of the present text, which deals with the evolution of single stars.

9.6 The luminosity of accretion and hard radiation sources

It is true that throughout our discussion of the evolution of stars we have excluded stellar interactions. These make up a field of study – a theory – in itself. However, many astronomical phenomena result from the fact that single stars are not found in a void (as we have assumed at the outset) but in a medium of low-density material. This text would not be complete unless we devote some thought to the effect this medium might have on the evolution of the embedded star, disregarding the source of the material, which does involve stellar interactions. Obviously, the star would accrete some of the surrounding material. Having known the outcome of mass ejection, we should not be surprised to find that the opposite effect of mass accretion has its own significant consequences.

When a star of mass M and radius R accretes an amount of mass δm coming from infinity, its (negative) gravitational potential energy decreases by an amount

$$\delta E_{\text{grav}} = \frac{GM\delta m}{R}.$$

9.6 The luminosity of accretion and hard radiation sources

If the material is accreted over a time interval δt, the average rate of gravitational energy release $\dot{E}_{\mathrm{grav}} = \delta E_{\mathrm{grav}}/\delta t$ is proportional to the average *accretion rate*, $\dot{M} = \delta m/\delta t$:

$$\dot{E}_{\mathrm{grav}} = \frac{GM\dot{M}}{R}. \tag{9.10}$$

If the star is to maintain thermal equilibrium, this energy surplus must be radiated away. Thus an *accretion luminosity* may be defined in relation to the accretion process:

$$L_{\mathrm{acc}} = \dot{E}_{\mathrm{grav}} = \frac{GM\dot{M}}{R}. \tag{9.11}$$

Obviously, the larger the gravitational field of a star, the larger would be its accretion luminosity. Thus, for example, if a star of $1 M_\odot$ were to double its mass, say, during 10^{10} yr (comparable to the age of the universe), it should accrete at an average rate $\dot{M} = 10^{-10} M_\odot$ yr^{-1}. A main-sequence star would thus produce a luminosity of $GM_\odot \dot{M}/R_\odot \approx 3 \times 10^{-3} L_\odot$, entirely *negligible* compared with the natural luminosity of such a star. For a white dwarf, the resulting luminosity would be about a hundred times higher, a few tenths $L_\odot$, *significantly higher* than the typical luminosity of white dwarfs. For a neutron star, it would reach $100 L_\odot$, while for a black hole (assuming the accretion radius to be R_{Sch}) it would approach $1000 L_\odot$.

The accretion rate is limited by the requirement that the resulting luminosity be lower than the Eddington critical luminosity. Otherwise, the radiation pressure exerted on the infalling material would push it back and prevent it from accumulating. We recall that the luminosity approaches the critical limit as the radiation pressure becomes dominant, and the binding energy of the star tends to zero. The requirement $L_{\mathrm{acc}} < L_{\mathrm{Edd}}$ leads, according to equation (5.37), to

$$\dot{M} < \frac{4\pi c R}{\kappa}, \tag{9.12}$$

which for electron scattering opacity translates into

$$\dot{M} < 10^{-3} R/R_\odot \ M_\odot \ \mathrm{yr}^{-1}. \tag{9.13}$$

We note that the upper limit of the accretion rate depends solely on the stellar radius, regardless of the mass.

In conclusion, even a moderate accretion rate (far below the upper limit) may induce the three types of compact stars to emit a significant luminosity. What kind of radiation would we expect in such instances? In order to maintain thermal equilibrium, by emitting the surplus gravitational energy, a star must adjust its surface temperature. The gravitational energy of the infalling matter is absorbed by a surface boundary layer, which acquires a temperature T_b and re-emits the energy as blackbody radiation. Since compact objects are stiff, the radius is barely affected. Hence T_b can be estimated by

$$T_b = \left(\frac{L_{\text{acc}}}{4\pi R^2 \sigma}\right)^{1/4} = \left(\frac{GM\dot{M}}{4\pi R^3 \sigma}\right)^{1/4}. \qquad (9.14)$$

An upper limit is obtained by substituting the critical value of $\dot{M}$ (9.13) on the right-hand side, which provides a reasonable estimate in view of the weak dependence of T_b on $\dot{M}$ (a power of $1/4$). For a white dwarf we obtain $T_b \approx 10^6$ K, for a neutron star $T_b \approx 1.5 \times 10^7$ K, and for a black hole, $T_b \gtrsim 3 \times 10^7$ K. These would appear as bright UV, X-ray, and even γ-ray sources. Indeed, such sources are quite abundant in our Galaxy (and beyond), as satellite-mounted modern detectors reveal.

Thus extinct compact stars, which would otherwise escape observation, may be rejuvenated by accretion. In fact, accretion leads to a wide variety of fascinating phenomena – an entire zoo of exotic objects – but the simple principles of stellar evolution that we have uncovered remain the same and can be applied to explain, often quite successfully, the evolution of binary stars, as they have explained the more straightforward evolution of single ones.

10

The stellar life cycle

10.1 The interstellar medium

Although to all intents and purposes a single star may be regarded as evolving isolated in empty space, not only is it member of a very large system of stars – a galaxy – but it is also immersed in a medium of gas and dust, the *interstellar medium*. This background material (mostly gas) amounts, in our Galaxy, to a few percent of the galactic mass, some $10^9 M_\odot$, concentrated in a very thin disc, less than 10^3 light-years in thickness (we recall that 1 ly $\simeq 9.5 \times 10^{15}$ m), and $\sim 10^5$ light-years in diameter, near the galactic midplane. Its average density is extremely small, about one particle per cubic centimetre, corresponding to a mass density of 10^{-21} kg m^{-3} (10^{-24} g cm^{-3}); in an ordinary laboratory it would be considered a perfect "vacuum." The predominant component of galactic gas – of which stars are formed – is hydrogen, amounting to about 70% of the mass, either in molecular form (H_2), or as neutral (atomic) gas (H I), or as ionized gas (H II), depending on the prevailing temperature and density. Most of the remaining mass is made up of helium. The interstellar material is not uniformly dispersed, but resides in clouds of gas and dust, also known as *nebulae*. We have already encountered special kinds of such nebulae: planetary nebulae and supernova remnants. These expanding nebulae are, however, relatively short-lived, and after dissipating into the interstellar medium, their material mixes with other, larger ones. There are relatively dense clouds, with number densities reaching up to a few thousand particles per cubic centimetre, and there is a diffuse intercloud medium, where densities can be much lower than one particle per cubic centimetre. The interstellar medium is extremely rich and diverse, which makes its exploration all the more fascinating.

When we speak of temperature in interstellar medium, we refer to the kinetic temperature of the gas. The radiation that fills the medium, emitted by the vast number of stars within it, is *not* in equilibrium with the gas, as it is in the stellar interiors. Nevertheless, it is this radiation that determines the *gas* temperature. The UV photons ionize the hydrogen atoms and the resulting free electrons collide with the ions. Although the mean free path of particles in the interstellar medium is about 10^{13} m, comparable to the diameter of the entire solar system, this amounts to only $\sim 10^{-3}$ light-years, a minute fraction of the typical cloud dimensions of tens to hundreds of light-years. Hence thermodynamic equilibrium is indeed achieved for the gas, and temperature is a meaningful concept.

Partly ionized gas clouds surrounding hot stars (such as massive main-sequence stars) may attain temperatures of the order of 10^4 K over regions of tens of light-years. The extent of such a region is obtained by requiring ionization balance: the number of absorbed ionizing photons must be equal to the number of recombinations, per unit volume per unit time. The H I zones of the interstellar medium (identified by the detection of the famous 21-cm radio line emitted by atomic hydrogen) have temperatures of 50–100 K. Roughly, the pressures within the different types of clouds are comparable (it is possible that the cold clouds, which are not gravitationally bound, are held together against their internal pressure by the hot gas component of the interstellar medium, which exerts a counterpressure). Hence the densities are in inverse proportion to the temperature. Typical number densities are $\sim 10^7$–10^8 m^{-3} for the cold clouds and $\sim 10^5$ m^{-3} for the hot gas. Besides the cold and hot clouds of neutral and ionized hydrogen, there are giant, dense, and dust-rich molecular clouds, where temperatures can be as low as 10 K, and number densities are in the range 1–3×10^8 m^{-3} and more. Their masses may reach $10^6 M_\odot$ and their sizes are of the order of 100 light-years. It is in these giant gaseous clouds that stars are born.

10.2 Star formation

The process of star formation constitutes one of the problems at the frontier of modern theoretical astrophysics. We shall not deal with the complicated stages that turn a fragment of an interstellar cloud into a star, but only address the question of the basic phenomenon of fragmentation.

Interstellar gaseous clouds are often subject to perturbations that are due, for example, to propagating shock waves originating in a nearby supernova explosion, or to collisions with other clouds. Consider an ideal case of a low-density cloud of uniform temperature T, in a state of hydrostatic equilibrium. If at some place a random perturbation will produce a region of higher density, the gravitational pull will increase in that region. The gas pressure will increase as well, but not necessarily in the amount required to maintain the hydrostatic equilibrium. The outcome of the perturbation will depend on the dynamical stability of that region. Our purpose is to derive the condition for stability for a region of volume V (which, for simplicity, may be assumed spherical), containing a given mass M. Denoting the radius (characteristic length) by R, we may use the *partial* virial theorem [equation (2.24) of Section 2.4], as we did in Section 8.4,

10.2 Star formation

to obtain

$$\int P\,dV = P_s V + \tfrac{1}{3}\alpha \frac{GM^2}{R}, \tag{10.1}$$

where P_s is the pressure at the region's boundary, exerted by the surrounding gas, and α is a constant of the order of unity (depending on the mass distribution within the region considered). We may assume an ideal gas equation of state [equation (3.28)], which yields

$$\int P\,dV = \frac{\mathcal{R}}{\mu}T \int \rho\,dV = \frac{\mathcal{R}}{\mu}TM. \tag{10.2}$$

Combining equations (10.1) and (10.2), we obtain

$$\frac{\mathcal{R}}{\mu}TM = P_s V + \tfrac{1}{3}\alpha \frac{GM^2}{R}. \tag{10.3}$$

Now, both P_s and V are positive quantities, and hence, obviously, the left-hand side of equation (10.3) must exceed the second term on the right-hand side, which means

$$R \geq \frac{\alpha}{3}\frac{\mu GM}{\mathcal{R}T}.$$

Equality is obtained when the entire cloud is involved. A critical radius (dimension) R_J may thus be defined by

$$R_J = \frac{\alpha}{3}\frac{\mu GM}{\mathcal{R}T}, \tag{10.4}$$

known as the *Jeans radius*, after Sir James H. Jeans, who was the first to investigate instabilities of this kind (in 1902). It constitutes a *lower limit for the dimension* of a stable region of temperature T, containing a given mass M, within a gaseous cloud. Contraction below this limit will cause the perturbed region to collapse: the gas pressure will be insufficient for balancing gravity. Conversely, we may obtain an *upper limit for the mass* that can be contained in hydrostatic equilibrium within a region of given volume, the *Jeans mass* M_J. With $\rho_{av} = M/V$,

$$M_J = \left[\left(\frac{3}{4\pi}\right)^{1/2}\left(\frac{3}{\alpha}\right)^{3/2}\right]\left(\frac{\mathcal{R}T}{\mu G}\right)^{3/2}\frac{1}{\sqrt{\rho_{av}}} \approx 10^5 \frac{T^{3/2}}{\sqrt{n}} M_\odot, \tag{10.5}$$

where n is the number of gas particles per m^3.

Inserting in equation (10.5) characteristic values of T and n for galactic gaseous nebulae, the Jeans mass turns out to be of the order of thousands to tens of thousands of solar masses, typical of stellar *clusters* rather than individual stars. Ordinary interstellar clouds have masses below this limit and hence they are stable. Only giant gas and dust complexes are prone to collapse. When collapse on such a scale is triggered, the question is how will it develop, and whether it will eventually stop. This is one of the crucial questions of the star formation theory.

Consider a collapsing cloud: both the density and the temperature increase and hence the value of the critical mass is expected to change. If the Jeans mass increases (inefficient cooling), we are faced with two possibilities: either the increase in M_J is sufficient for the stability criterion to be satisfied, in which case the collapse will halt, or M_J is still smaller than the cloud's mass, in which case the collapse will continue. If, on the other hand, M_J decreases (efficient cooling), the violation of the stability criterion is yet more severe; it may now happen that regions *within* the cloud violate the stability criterion and start collapsing, inducing fragmentation of the cloud. The fragmentation process may go on to smaller and smaller scales, down to the stellar mass scale. Such a hierarchical model was first suggested by Hoyle in 1953. Which of the possible situations will actually occur depends on the ratio between the timescale of collapse, which is the dynamical timescale of the cloud (of the order of $1/\sqrt{G\rho_{av}}$), and the cooling (thermal) timescale. Since cloud densities are many orders of magnitude lower than those prevailing in stars, these timescales are comparable and hence an accurate evaluation of the processes involved in the collapse is required. As cloud fragments become increasingly denser and hotter, they eventually become opaque and cooling becomes inefficient. At some point, the Jeans mass starts increasing. Thus, depending on local conditions, a *minimum Jeans mass* exists, which defines a lower limit to fragments of clouds that are bound to contract and form stars. A schematic illustration of fragmentation is given in Figure 10.1, and observational evidence for the process of collapse and fragmentation is shown in Figure 10.2.

> **Exercise 10.1:** Estimate the minimum Jeans mass of a collapsing isothermal gas cloud of temperature T, on the assumption that the radiation temperature is lower than the gas temperature (since there is not sufficient time for thermodynamic equilibrium to be achieved).

A fragment of a gas cloud bound by self-gravity, which has a mass in the stellar mass range, may be regarded as a nucleus of a future star. The mass continues to grow by accretion of gas from the surroundings. The gravitational energy released as material accretes is turned into thermal energy. The increase in both density and temperature raises the opacity of the gas. When the contracting gas

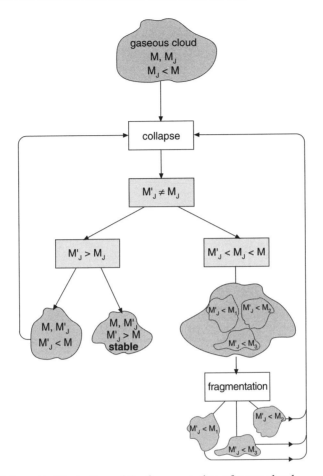

Figure 10.1 Schematic illustration of the fragmentation of a gas cloud.

becomes opaque to its own radiation, it has reached the status of a stellar embryo, the photosphere defining the boundary between the inside and the outside of the star in the making. When hydrostatic equilibrium is achieved, the embryo becomes a protostar (see Section 8.1). Eventually, the central temperature reaches the hydrogen ignition threshold and the protostar becomes a star, assuming its place on the main sequence of the H–R ($\log L$, $\log T_{\text{eff}}$) diagram appropriate to its mass.

10.3 Stars, brown dwarfs, and planets

The process of star formation has nothing to do with the ability of a star to ignite hydrogen when the turbulent stages leading up to ignition are finally over. Hence we cannot grant the protostellar cloud the prescience of having to end up with a mass above the lower stellar mass limit of about 0.08 $M_\odot$. Indeed, the estimated minimum Jeans mass is about an order of magnitude lower than the

Figure 10.2 Observational evidence of fragmentation in a gaseous cloud: starbirth clouds and gas pillars in the Eagle Nebula (M16), a star forming region 7000 ly away. The tallest pillar (left) is about 1 ly long from base to tip. Note the small globules of denser gas buried within the pillars [photograph by J. Hester and P. Scowen, Arizona State University, taken with NASA's Hubble Space Telescope].

lower stellar mass limit. Therefore, smaller objects should be expected to form by the same process that creates stars, only to start cooling before they could ignite hydrogen. Such objects have been observed, or their existence has been indirectly inferred from its effect on a binary companion. They are called *brown dwarfs*, to be distinguished from the common, bright *white dwarfs*, which will eventually become extinct *black dwarfs*, and from the lower main-sequence stars that are often referred to as *red dwarfs*, due to their reddish colour, resembling that of red giants. In the H–R diagram brown dwarfs descend the Hayashi track, but they turn away from the main sequence toward lower effective temperatures. In the ($\log T_c$, $\log \rho_c$) diagram, they start by contracting and heating up, as stars do, but their tracks bend into the degeneracy zone before crossing the hydrogen burning threshold. Subsequently, they behave much in the same way as giant planets. Planets, however, form in a different way: they separate out of circumstellar discs surrounding very young stars, by aggregation of larger and larger particles and by accretion of gas.

Hence brown dwarfs constitute a transitional class of objects, between stars and planets: they are born like stars, but they evolve like planets. In fact, they may

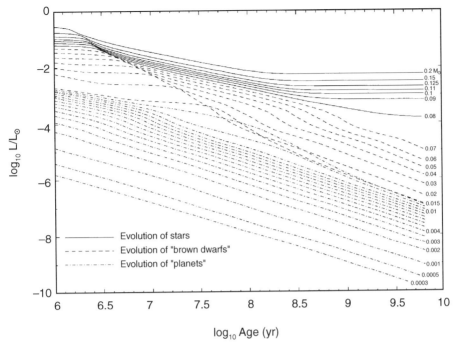

Figure 10.3 Evolution of the luminosity of red dwarf stars (solid curves), brown dwarfs (dashed curves), and planets (dash-dotted curves). Brown dwarfs are here identified as those objects that burn deuterium. Curves are labelled according to mass, the lowest three corresponding to the mass of Jupiter, then half of Jupiter's mass, and finally the mass of Saturn [from A. Burrows et al. (1997), *Astrophys. J.*, 491].

even have a claim to stardom, since they do briefly ignite deuterium (primordial deuterium being present in very small amounts, of the order of 10^{-5}, in the initial composition of all stars). The evolution of the luminosity for objects in the mass range $0.0003–0.2 M_\odot$ resulting from model calculations is shown in Figure 10.3. The early flat part of the tracks, between 10^6 and 10^8 yr, is due to deuterium burning; this phase is very short in the more massive stars, but can last as long as 10^8 yr in an object of $\sim 0.01 M_\odot$, at the lower mass limit for deuterium burning. After about 10^8 yr the stars among these objects reach a plateau luminosity upon settling on the main sequence. For planets, by contrast, the luminosity decreases continuously. Brown dwarfs fall in between, with a brief period of constant luminosity, followed by a steady decline.

Thus another distinction may be made between brown dwarfs and planets, not according to birth, but according to whether or not they have ever burnt nuclear fuel. Strangely enough, both definitions – although having nothing in common – result in similar lower limits for brown dwarf masses, $0.01 \pm 0.003 M_\odot$. Yet a further distinction may be made according to structure. In very low mass stars and brown dwarfs the internal pressure is supplied mainly by the degeneracy pressure

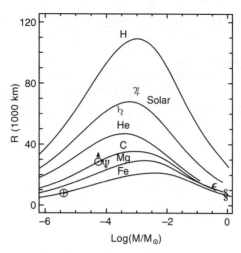

Figure 10.4 Mass-radius relation for low-mass objects [following H. S. Zapolsky & E. E. Salpeter (1969), *Astrophys. J.*, 158]. Different curves correspond to different compositions, as indicated. The locations of several planets – Earth, Jupiter, Saturn, Uranus, and Neptune – are marked by the planets' symbols. Also marked are the locations of two white dwarfs – Sirius B (§) and 40 Eridani B (ϵ) [data from D. Koester (1987), *Astrophys. J.*, 322].

of electrons, similarly to white dwarfs, except that white dwarfs are much closer to complete degeneracy and hence, in a way, simpler to model, and they are made of elements heavier than hydrogen. We have seen that for objects dominated by degeneracy pressure radii increase with decreasing mass (Section 5.4). But such behaviour cannot go on indefinitely. We know, for example, that for terrestrial planets, which are governed by much more complicated equations of state, radii decrease with the mass. Therefore, a mass must exist for which the radius, as a function of mass, reaches a maximum. The mass-radius relation for spheres of low mass based on an accurate equation of state is shown in Figure 10.4 for different compositions. As it turns out, the mass corresponding to the maximal radius is very close to Jupiter's. Hence Jupiter's mass, $M_J \approx 0.001 M_\odot$, may be regarded as a borderline between two classes of objects. Indeed, brown dwarf masses are often expressed in units of M_J, ranging from about $80 M_J$ down to about $10 M_J$ (or less?).

However, none of the criteria mentioned above for distinguishing brown dwarfs from planets can be applied observationally; they are all based on history or internal structure. In order to identify brown dwarfs we need to specify surface characteristics, such as spectral signatures. These are difficult to determine because opacities at low temperatures are complicated by the formation of molecules and dust grains. In fact, the interest in these small and faint objects has been aroused by their kinship to planets, which are currently at the focus of astronomical research, in the attempt to answer the intriguing question of extraterrestrial life.

We have yet a great deal to learn about the nature of brown dwarfs, about giant planets, and about the formation of stars and planets, until we shall be able to sort out and fully understand the variety of substellar objects, even before we address the question of the origin of life. Another reason for the increasing interest in brown dwarfs is their potential contribution to the galactic mass budget in the form of "dark" matter. For this contribution to be significant, their number must be considerable. This brings us to the question of the stellar mass distribution.

> **Note:** Dark matter is matter that we do not see (in any wavelength), but we have other indications to presume it is there. These come mainly from the gravitational field that such matter would generate, just as in the case of black holes. On the galactic scale, the evidence is provided by fast-moving stars and gas clouds at the very edge of the revolving galactic disc, where Keplerian velocities should be much smaller, if the gravitational field were due to visible matter alone. At such velocities these stars and clouds should have long dispersed, unless pulled in by the gravitational field of an invisible material halo. On larger scales, a similar phenomenon is observed in clusters of galaxies, as was pointed out by Zwicky in the 1930s. The random motions of galaxies within a cluster tend to disperse it, while the mutual gravitational pull would cause them to fall to the centre. Thus balance is established, with the random velocities being related to the cluster's mass (as in the virial theorem that applies to a self-gravitating gas; Section 2.4). As it turns out, the observed velocities (deduced from Doppler shifts) of cluster members exceed by far those that correspond to the visible mass. In order to keep them confined to the cluster, a mass exceeding their own by a factor of almost ten would be required. Hence the quest for "dark" matter.

10.4 The initial mass function

Continual star formation results in a steady decrease in the population of massive, luminous stars. As these stars have vanishingly short life spans on the galactic timescale, their relative number is at each instant correlated with the fractional amount of gas in a galaxy. Thus even if they were created with the same probability as low-mass stars, massive stars would have become rarer in the course of galactic evolution. All the more so, if the probability of formation of massive stars is small relative to that of low-mass stars, as turns out to be the case.

Assuming star formation to be independent of galactic age or location, the number of stars formed at a given time within a given volume, with masses in the range $(M, M + dM)$, is solely a function of M:

$$dN = \Phi(M)dM. \tag{10.6}$$

The so-called *birth function* $\Phi(M)$ was derived by Salpeter as early as 1955 and it has hardly changed since:

$$\Phi(M) \propto M^{-2.35}. \tag{10.7}$$

The related *initial mass function* $\xi(M)$ is defined as follows: the amount of mass locked up in stars with masses in the interval $(M, M + dM)$, formed at a given time within a given volume, is

$$M\,dN = \xi(M)\,dM, \tag{10.8}$$

and combining relations (10.6)–(10.8), we have

$$\xi(M) \propto \left(\frac{M}{M_\odot}\right)^{-1.35}. \tag{10.9}$$

The semi-empirical derivation of relation (10.7) was based on observations of main-sequence star luminosities in the solar neighbourhood. It involved the division of the luminosity range of main-sequence stars into intervals, counting the total number of stars with luminosities in each interval, using the mass-luminosity relation (1.6), and, finally, assuming that the duration of the main-sequence phase was proportional to M/L. This assumption implies that stars leave the main sequence as soon as they have burnt a fixed fraction of their mass, as indicated by the stellar evolution theory. In spite of the enormous increase in observational data since the early 1950s and the refinements of stellar evolution theory, the conclusion remains that over a mass range spanning more than two orders of magnitude, from $\sim 0.3 M_\odot$ to $\sim 60 M_\odot$, the birth function is of the form $\Phi(M) \propto M^{-1-\gamma}$, with $\gamma \approx 1.5 \pm 0.3$. An example based on main-sequence stars of the galactic disc population around the Sun is given in Figure 10.5. There are many theories that try to explain this empirical birth function, but so far none has been generally accepted.

We note that at the low-mass end, the initial mass function deviates considerably from the inverse power law (10.9) and becomes almost flat and even decreasing with mass. The difficulties involved in observing the faint low-mass stars and brown dwarfs and obtaining complete samples make the derivation of a birth function in this range rather uncertain. However, it is already clear that the total mass of objects with $M \lesssim 0.3 M_\odot$ can account for less than 20% of the total mass of stars with $M \gtrsim 0.3 M_\odot$. Thus the solution to the missing mass problem should probably be sought elsewhere. We should also mention that recent observations indicate a conspicuous change in slope for the initial mass function around the transition mass between brown dwarfs and planets. This strengthens

10.4 The initial mass function

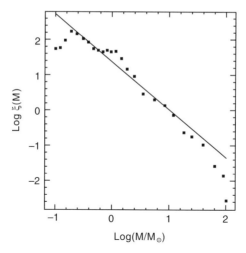

Figure 10.5 The initial mass function of main-sequence stars in the solar neighbourhood. The Salpeter slope is indicated by the straight line [data from N. C. Rana (1987), *Astron. Astrophys.*, 184].

the hypothesis that these two types of objects were formed by radically different processes.

With the aid of the initial mass function, rough estimates may be derived for the mass exchange between stars and their environment, and for stellar distributions within a volume of the galaxy. Consider, for example, one generation of stars formed at a given time in some part of the galaxy. The fractional amount of mass returned to the galactic medium by this generation of stars may be computed as follows. Let ζ be the mass initially locked up in these stars, whose masses are in the range $M_{\min} < M < M_{\max}$. Then

$$\zeta = \int_{M_{\min}}^{M_{\max}} M\,dN = \int_{M_{\min}}^{M_{\max}} \xi(M)\,dM. \tag{10.10}$$

Let $\mathfrak{R}$ denote the fraction of initial mass that a star ejects in the course of its evolution. As we have seen in Chapter 9, stars that end their lives in a supernova explosion eject more than 80% of their mass. For the purpose of a rough estimate, we may assume that stars of initial mass above $M_{\rm SN} \gtrsim 10 M_\odot$ return their entire mass to the galactic medium. Stars of initial mass below $M_{\rm MS} \approx 0.7 M_\odot$ will still be in the main-sequence phase, as we have seen in Section 8.2. These stars have lost, therefore, only a negligible fraction of their initial mass. Stars in the intermediate range $M_{\rm MS} < M < M_{\rm SN}$ may be taken to have turned instantaneously into white dwarfs, since the time elapsed between the main-sequence phase and the white dwarf phase is relatively short (see Sections 8.4–8.7). These stars have thus ejected all but the remnant white dwarf's mass ($M_{\rm WD} \sim 0.6 M_\odot$).

Consequently,

$$\mathcal{R} \approx \begin{cases} 1 & ; & M \geq M_{SN} \\ (M - M_{WD})/M & ; & M_{MS} < M < M_{SN} \\ 0 & ; & M \leq M_{MS} \end{cases}, \qquad (10.11)$$

and the mass returned by this generation of stars to the interstellar medium is

$$\eta = \int_{M_{\min}}^{M_{\max}} \xi(M)\mathcal{R}(M)\,dM = \int_{M_{MS}}^{M_{\max}} \xi(M)\,dM - \int_{M_{MS}}^{M_{SN}} \frac{M_{WD}}{M} \xi(M)\,dM. \qquad (10.12)$$

The fractional mass returned is obtained by dividing equation (10.12) by equation (10.10),

$$\frac{\eta}{\zeta} = \frac{\int_{M_{MS}}^{M_{\max}} M^{-1.35}\,dM - M_{WD} \int_{M_{MS}}^{M_{SN}} M^{-2.35}\,dM}{\int_{M_{\min}}^{M_{\max}} M^{-1.35}\,dM}, \qquad (10.13)$$

which amounts to $\sim 1/3$, for $M_{\min} = 0.1 M_\odot$ and $M_{\max} = 60 M_\odot$.

Exercise 10.2: Test the sensitivity of the above estimate for η/ζ to the stellar mass range assumed, by repeating the calculations for all combinations of $M_{\min} = 0.05$ and $0.2 M_\odot$, and $M_{\max} = 30$ and $120 M_\odot$.

We may also estimate the number of white dwarfs relative to the number of main-sequence stars in a population of stars formed at a given time, such as a stellar cluster. All we need to know is the mass corresponding to the upper end of the main sequence in the H–R diagram of the cluster – the mass of the *turn-off point* M_{tp}. The number of main sequence stars is then given by

$$N_{MS} = \int_{M_{\min}}^{M_{tp}} dN = \int_{M_{\min}}^{M_{tp}} \Phi(M)\,dM. \qquad (10.14)$$

The number of white dwarfs is obtained by assuming, as before, that the transition from the end of the main-sequence state to the white dwarf state is instantaneous:

$$N_{WD} = \int_{M_{tp}}^{M_{SN}} dN = \int_{M_{tp}}^{M_{SN}} \Phi(M)\,dM. \qquad (10.15)$$

10.4 The initial mass function

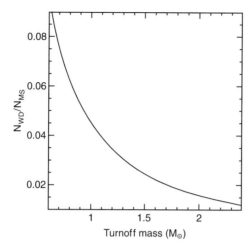

Figure 10.6 Ratio of number of white dwarfs to number of main-sequence stars for a stellar ensemble of given age, where the age is given by the mass corresponding to the turning point off the main sequence in the H–R diagram (cf. Figure 8.5).

Hence the ratio

$$\frac{N_{WD}}{N_{MS}} = \frac{M_{tp}^{-1.35} - M_{SN}^{-1.35}}{M_{min}^{-1.35} - M_{tp}^{-1.35}}, \qquad (10.16)$$

which amounts to only a few percent, is a function of M_{tp}, or the cluster's age, as shown in Figure 10.6.

> **Exercise 10.3:** Assume the brightness of a stellar cluster to be mainly determined by the summed luminosities of its main-sequence stars. Calculate by what factor would a cluster's brightness decrease, as the turnoff point of its main sequence moves down from $1.3 M_\odot$ to $0.85 M_\odot$.

A similar estimate of the number of supernovae (or, equivalently, neutron stars) relative to that of main-sequence stars yields a lower ratio by more than a factor of 10. We should keep in mind, however, that supernovae would be visible only during the early evolution of the cluster, up to about the main-sequence life span of a star of mass M_{SN}, and if no further star formation occurs, no stellar explosions should be seen thereafter. Stellar statistics, which take into account both variations due to stellar masses *and* variations due to different ages, involve far more complicated calculations and represent a separate field of study – a very important one, considering that many of the tests of the stellar evolution theory are of a statistical nature.

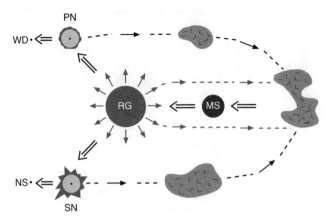

Figure 10.7 Sketch of the stellar evolution cycle.

10.5 The global stellar evolution cycle

On large scales, the process of stellar evolution is a cyclic process: stars are born out of gaseous clouds within galaxies, and in the course of their lives they return to the galactic medium a large fraction of the mass they have temporarily trapped. This material blends with the interstellar matter and contributes, in turn, to the formation of new generations of stars. This is sketched in Figure 10.7. The term "generation of stars" is somewhat misleading, for we have seen that stellar lifetimes differ by as much as four orders of magnitude, depending on the initial mass. Thus a succession of a great many generations of massive stars may coincide with only one single generation of low-mass stars. The different ways by which stars return material to the interstellar medium are illustrated by the images of Figure 10.8, where the shell ejected by a nova outburst (*novae* have been briefly mentioned in Section 6.2) is shown in addition to the more familiar wind from a massive star, another example of a planetary nebula, and the shell ejected by supernova SN1987A (Section 9.3). We note the conspicuous similarity of these images, despite the huge differences in length and time scales. The ejected material has been processed, however, and its composition differs from the prevailing composition of the galactic gas. Thus, later generations of stars have, at birth, increasingly larger abundances of heavy elements (or *metals*). The survivors of the entire evolution process are dense compact stars – white dwarfs, neutron stars, and, possibly, black holes – as well as brown dwarfs and low-mass main-sequence stars, whose main sequence life spans exceed the age of the universe. In the end, when the entire gas reservoir will have been locked up in these small and mostly faint stars, star formation will cease.

The main evolutionary processes that take place on the galactic scale as a result of individual stellar evolution may be summarized as follows.

10.5 The global stellar evolution cycle

Figure 10.8 Illustration of mass loss by images taken with NASA's Hubble Space Telescope: *top left*: nebula ejected by a massive star (estimated at $\sim 100 M_\odot$) extending in radius to ~ 4 ly [photograph by D. F. Figer, University of California at Los Angeles]; *top right*: mass ejected by SN1987A: the ring of gas, about 1.5 ly in diameter, was expelled by the progenitor star some 2×10^4 yr before the supernova explosion. At its centre, the glowing gas ejected in the explosion expands at a speed of 3000 km s^{-1} [photograph by P. Garnavich, Harvard-Smithsonian CFA]; *bottom left*: planetary nebula (Henize 1357), the youngest known so far, extending to a radius of less than one tenth ly [photograph by M. Bobrowsky, Orbital Science Corp.]; *bottom right*: mass shells ejected by nova T Pyxidis, forming more than 2000 gaseous blobs, which extend to a diameter of about 1 ly [photograph by M. Shara, R. Williams, and D. Zurek, Space Telescope Science Institute; R. Gilmozzi, European Southern Observatory; and D. Prialnik, Tel Aviv University].

1. The amount of free gas decreases. Nebulae and gas clouds become sparse.
2. The galactic luminosity – made up of the individual stellar luminosities – declines, as the relative number of massive stars decreases at the expense of the growing proportion of compact, faint stars.

3. The composition becomes enriched in heavy elements, created in stars and returned to the galaxy by the various processes of mass ejection.

Exercise 10.4: Let $\Upsilon(t)$ be the fractional amount of gas in the Galaxy as a function of time, satisfying the initial condition $\Upsilon(0) = 1$. Assume the rate of decrease of free gas as a result of star formation to be proportional to Υ^2. Find the function $\Upsilon(t)$, if at present, $t = t_p$, the gas constitutes 0.05 of the galactic mass. At what time in the past (fraction of t_p) was the mass of free gas half the entire mass? At what time was it one tenth of the entire mass? At what future time will the gas mass have decreased to half its present value?

As galactic material is continually enriched in heavy elements, the relative abundance of these elements in newly formed stars increases with time elapsed from the formation of the galaxy. Despite their relatively low birth rate, it is the massive stars that provide the overwhelming contribution to heavy element enrichment, first because they eject a larger fraction of their initial mass than do low-mass stars, secondly because this fraction constitutes a much larger amount of mass, and thirdly because the mass is returned practically instantaneously. The contributions of stars of different masses to interstellar helium and metal enrichment by different processes are illustrated in Figure 10.9. These yields have to be weighted by the initial mass function to properly derive the galactic chemical evolution. The rate of heavy element enrichment of the galactic medium has been far from constant. Most of the enrichment occurred at early times, and the enrichment rate has markedly decreased with time. As an illustration, we note that the age of the Sun is about one third of the age of the Galaxy, and its heavy element mass fraction (metallicity) Z is nearly 0.02. The metallicity of the youngest stars is about 0.04, that of the oldest, about 0.0003. Thus Z has increased a hundredfold during the first two thirds of the galactic lifetime and only twofold during the last third.

Although the change in initial abundance is gradual, it has become customary to divide stars into two populations, *Population I* (Pop I, for short) and *Population II* (Pop II), according to composition and hence to age. The stars of Pop I are young and *metal* rich; those of Pop II are old and *metal* poor. If we reverse the time arrow from the present backward, into the past, we first encounter the Pop I stars and then those belonging to Pop II. This could be taken as the rationale for ordering the populations. Thus *old* Pop I stars are those stars formed in between Pop I and Pop II. And sometimes reference to *Population III* stars may be found, meaning that we have to go further down the time arrow, passing the *extreme* Pop II stars, toward the very beginning of galactic evolution. On this time arrow Z decreases, with older populations corresponding to lower Z values.

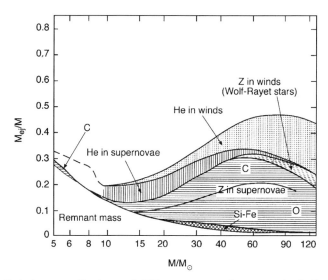

Figure 10.9 Relative contributions of different types of stars to the helium and heavy element content of the interstellar medium [adapted from C. Chiosi & A. Maeder (1986), *Ann. Rev. Astron. Astrophys.*, 24].

However, the initial classification of stars into distinct populations, dating back to 1944, when it was introduced by Baade, was based neither on initial composition nor on age, but on the location of stars in the galaxy. A typical spiral galaxy, such as the Milky Way, consists of a spherical distribution of stars which includes the central *bulge* and the galactic *halo*, and a flattened disc distribution, which is not uniform, but divides into spiral arms. The disc stars made up Pop I, while the stars in the central region of the galaxy and in the globular clusters forming the galactic halo constituted Pop II. Location, age, and composition are thus interconnected. The galactic disc, which contains most of the free gas and dust, is still harbouring star formation. Hence, not surprisingly, young stars are abundant in the disc. The halo is a relic of the original distribution of matter in the galaxy, before most of the material collapsed to form the disc. It contains the old high-velocity stars and globular clusters that formed before the collapse of the disc and retained their high kinetic energy. The H–R diagrams of the different populations are consistent with the inference of their relative ages.

But even in the youngest stars the *mass fraction* of heavy elements amounts to only a few percent. On the one hand, this is too little to affect a star's evolution considerably, although the evolutionary paths traced by stars of different metallicities in the H–R diagram do differ to some extent. On the other hand, the total *absolute mass* of heavy elements in a star is rather considerable; in the Sun, for example, it exceeds the mass of the entire solar system, including planets, moons, comets, asteroids, and other star formation debris. From where we stand this cannot be considered negligible. In fact, except for the giant planets, which contain a significant amount of (primordial) gas, all the other bodies in the solar system are made precisely of some of that small fraction of heavy elements present in the protosolar nebula. And, as we recall that the source of these elements has been nuclear burning, we come to the awesome conclusion that most atoms in our bodies, the atoms in the air that we breathe, and, in short, the elements making up every object around us, have belonged to a star at some time in the past and, in all probability, have witnessed a gigantic stellar explosion.

. . .

And steadfast as Keats' Eremite,
Not even stooping from its sphere,
It asks a little of us here.
It asks of us a certain height,
So when at times the mob is swayed
To carry praise or blame too far,
We may choose something like a star
To stay our eyes on and be staid.

Robert Frost

Appendix 1

The equation of radiative transfer

Consider a small cylinder of length $d\ell$ and cross-section dS at a distance r from the centre of a star, with its axis in the direction θ with respect to the radius vector. The projected length of the cylinder is thus $dr = d\ell \cos\theta$ (as shown in Figure A1.1). The radiation within the cylinder obeys energy conservation. We define the *intensity of radiation* $I(r, \theta)$ such that $I(r, \theta)d\omega$ is the energy flux (energy per unit area per unit time) moving inside a cone of directions defined by the solid angle $d\omega = 2\pi \sin\theta d\theta$ around the direction of θ. Since the radiation is composed of photons of different frequencies and since interactions between matter and radiation depend on frequency, we consider the monochromatic intensity $I_\nu(r, \theta)$ defined such that

$$I(r, \theta) = \int_0^\infty I_\nu(r, \theta) d\nu, \qquad (A1.1)$$

and we apply energy conservation in each frequency.

The different contributions to the radiation energy during a time interval dt are as follows.

1. The energy entering the cylinder at the bottom is

$$dQ_{\text{in}} = I_\nu(r, \theta) d\omega \, dS \, dt.$$

2. The energy leaving the cylinder at the top is

$$dQ_{\text{out}} = -I_\nu(r + dr, \theta) d\omega \, dS \, dt.$$

In fact, along the cylinder, the angle between the axis and the radial direction decreases and there should be a difference $d\theta$ between the top and the bottom. For simplicity, we neglect this difference, which is tantamount to adopting the plane parallel approximation. (In the general case, the same

Appendix I The equation of radiative transfer

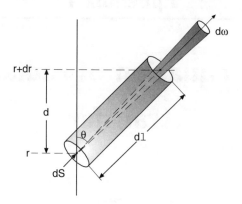

Figure A1.1 Cylindrical volume element within a star; conservation of radiation energy within it leads to the radiation transfer equation.

basic relations are reached as we shall obtain here, following the same line of reasoning, but the mathematics is a little more complicated.)

3. The absorbed radiation within the cylinder is

$$dQ_{abs} = -\kappa_\nu \rho I_\nu(r, \theta) d\omega \, dS \, d\ell \, dt.$$

We distinguish between true absorption and absorption caused by scattering, and we label the opacity coefficients accordingly; thus $\kappa_\nu = \kappa_{a,\nu} + \kappa_{s,\nu}$.

4. The emitted radiation by the mass within the cylinder is

$$dQ_{em} = \rho j_\nu d\omega \, dS \, d\ell \, dt,$$

where j_ν is the total radiation emitted per unit mass per unit time. We include in this term radiation emitted by the mass within the cylinder, $j_{em,\nu}$, as well as radiation scattered *into* the cylinder, $j_{s,\nu}$. The latter is obtained by integrating $\kappa_{s,\nu} I_\nu(r, \theta')$ over all directions θ' from which photons are scattered into our cylinder, assuming that the scattering process does not change the photon frequency. This is usually a complicated task. In the simple isotropic case, that is, when the scattered radiation is emitted equally into equal solid angles, $j_{s,\nu} = \kappa_{s,\nu} \int I'_\nu d\omega'/4\pi$.

Conservation of energy requires

$$\Sigma dQ = 0,$$

and hence

$$[-I_\nu(r+dr, \theta) + I_\nu(r, \theta) - \kappa_\nu \rho I_\nu(r, \theta) d\ell + \rho j_\nu d\ell] dS \, d\omega \, dt = 0. \quad (A1.2)$$

Appendix I The equation of radiative transfer

Subsituting

$$I_\nu(r+dr,\theta) - I_\nu(r,\theta) = \frac{\partial I_\nu(r,\theta)}{\partial r}dr = \frac{\partial I_\nu(r,\theta)}{\partial r}d\ell\cos\theta$$

and dividing equation (A1.2) by $d\ell\,dS\omega dt$, we obtain the radiative transfer equation in the form

$$\frac{1}{\rho}\frac{\partial I_\nu(r,\theta)}{\partial r}\cos\theta + \kappa_\nu I_\nu(r,\theta) - j_\nu = 0. \qquad (A1.3)$$

Note that the scattering term $j_{s,\nu}$ turns the transfer equation into an *integro-differential* equation. In order to solve it, we have to evaluate $j_{em,\nu}$, itself a function of I_ν.

In thermodynamic equilibrium, the radiation field is given by the Planck (blackbody) distribution

$$B_\nu(T) = \frac{2h\nu^3}{c^2}\frac{1}{e^{h\nu/kT}-1}, \qquad (A1.4)$$

which is isotropic, and there is perfect balance between absorption and emission of radiation (known as Kirchhoff's law); we then have $I_\nu(r) = B_\nu(T)$ and $j_{em,\nu} = \kappa_{a,\nu}B_\nu(T)$. In stars, however, the radiation field is not perfectly isotropic, and hence we have to consider the different contributions to the emission of radiation. It was Einstein who recognized that these must be of two kinds: *spontaneous* emission, determined by the temperature, and *induced* (or *stimulated*) emission, which is caused by the radiation field itself. The relationship between them and between emission and absorption may be easily understood by considering a simple case of two discrete energy levels 1 and 2, such that $E_2 = E_1 + h\nu$. Let n_1 and n_2 be the number densities of particles in the energy states E_1 and E_2, respectively. In thermodynamic equilibrium a second condition is satisfied: particle densities are related by Boltzmann's formula

$$\frac{n_2}{n_1} = \frac{g_2}{g_1}e^{-(E_2-E_1)/kT} = \frac{g_2}{g_1}e^{-h\nu/kT}, \qquad (A1.5)$$

where the factors $g_{1,2}$ represent the statistical weights of the energy states (essentially, the number of states with different quantum numbers that correspond to the same energy level). Transition of a particle from level 2 to level 1 involves the emission of a photon of energy $h\nu$; similarly, the reverse transition occurs by absorption of such a photon, as shown schematically in Figure A1.2. The rate of spontaneous emission is proportional to the number of particles in the high energy state n_2; the rate of induced emission, on the other hand, depends on both n_2 and the radiation field $B_\nu(T)$. Finally, the rate of absorption is proportional to the number of particles in the low energy state n_1 and to the radiation field. Introducing the appropriate coefficients – A_{21} for spontaneous emission, B_{21} for induced emission, and B_{12}

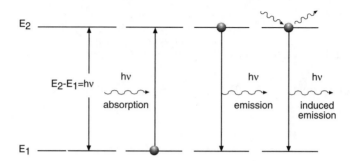

Figure A1.2 Schematic representation of emission and absorption in a two energy-level system.

for absorption – and applying Kirchhoff's law, we obtain Einstein's equation:

$$A_{21}n_2 + B_{21}n_2 B_\nu(T) = B_{12}n_1 B_\nu(T), \tag{A1.6}$$

where we identify on the right-hand side $B_{12}n_1 = \kappa_{a,\nu}$. Multiplying by $(e^{h\nu/kT}-1)$, defining $\alpha_\nu = 2h\nu^3/c^2$ and substituting n_2 from equation (A1.5), we have

$$g_2 A_{21}(e^{h\nu/kT} - 1) + g_2 B_{21}\alpha_\nu = g_1 B_{12}\alpha_\nu e^{h\nu/kT}. \tag{A1.7}$$

This equation holds for any temperature, regardless of photon frequency; hence temperature-dependent and temperature-independent terms must balance separately. We thus obtain the *Einstein relations* between the three coefficients:

$$g_2 A_{21} = g_1 B_{12}\alpha_\nu \tag{A1.8.1}$$

$$g_2 B_{21} = g_1 B_{12} \tag{A1.8.2}$$

(the second follows from the first and $A_{21} = B_{21}\alpha_\nu$), which leave only one independent coefficient.

Now comes the crucial point of the discussion: these relations must hold whether or not the system is in thermodynamic equilibrium. This is because they are connected to the microscopic state of the system – the nature of individual emitters-absorbers – whereas thermodynamic equilibrium is a macroscopic property. Individual particles are unaware, as it were, of the general state of the system. Consequently, for *any* radiation field intensity I_ν the emission is given by

$$j_{em,\nu} = A_{21}n_2 + B_{21}n_2 I_\nu = \frac{g_1}{g_2}B_{12}\alpha_\nu n_2 + \frac{g_1}{g_2}B_{12}n_2 I_\nu = \frac{g_1}{g_2}B_{12}n_1 \frac{n_2}{n_1}(\alpha_\nu + I_\nu). \tag{A1.9}$$

Substituting equations (A1.5) and (A1.4) into equation (A1.9) yields

$$j_{em,\nu} = \kappa_{a,\nu}(1 - e^{-h\nu/kT})B_\nu(T) + \kappa_{a,\nu}e^{-h\nu/kT}I_\nu, \tag{A1.10}$$

Appendix I The equation of radiative transfer

and it is easy to see that in the case of thermodynamic equilibrium this relation reduces to $j_{\text{em},\nu} = \kappa_{a,\nu} B_\nu(T)$, because then $I_\nu = B_\nu(T)$.

Defining a *reduced* absorption opacity coefficient by

$$\kappa_{a,\nu}^* = \kappa_{a,\nu}\left(1 - e^{-h\nu/kT}\right), \tag{A1.11}$$

and substituting expression (A1.10) into equation (A1.3), we obtain the transfer equation in the form

$$\frac{1}{\rho}\frac{\partial I_\nu(r,\theta)}{\partial r}\cos\theta + \kappa_{a,\nu}^*[I_\nu(r,\theta) - B_\nu(T)] + \kappa_{s,\nu} I_\nu(r,\theta) - j_{s,\nu} = 0, \tag{A1.12}$$

which can be solved for given opacity coefficients in all radiation frequencies.

To obtain a solution, we expand $I_\nu(r,\theta)$ in Legendre polynomials $P_n(\cos\theta)$,

$$I_\nu(r,\theta) = I_{\nu,0}(r)P_0 + I_{\nu,1}(r)P_1(\cos\theta) + I_{\nu,2}(r)P_2(\cos\theta) + \ldots, \tag{A1.13}$$

recalling that $P_0 = 1$ and $P_1 = \cos\theta$. Since matter and radiation are in local thermodynamic equilibrium in stars (see Section 2.1), we know that the first (isotropic) term in the expansion is none other than the Planck distribution $B_\nu(T)$. Substituting it into the transfer equation (A1.12) yields

$$\frac{1}{\rho}\frac{dB_\nu(T)}{dr}P_1 + \frac{1}{\rho}\frac{dI_{\nu,1}}{dr}\cos\theta\, P_1 + \ldots + \kappa_{a,\nu}^* I_{\nu,1} P_1 + \kappa_{a,\nu}^* I_{\nu,2} P_2 + \ldots$$
$$+ \kappa_{s,\nu} B_\nu(T) + \kappa_{s,\nu} I_{\nu,1} P_1 + \ldots - j_{s,\nu} = 0. \tag{A1.14}$$

We now use the recurrence relation of Legendre polynomials,

$$\cos\theta\, P_n = \frac{n}{2n+1}P_{n-1} + \frac{n+1}{2n+1}P_{n+1},$$

to obtain

$$\frac{1}{\rho}\frac{dB_\nu(T)}{dr}P_1 + \frac{1}{\rho}\sum_{n=1}^{\infty}\frac{dI_{\nu,n}}{dr}\left(\frac{n}{2n+1}P_{n-1} + \frac{n+1}{2n+1}P_{n+1}\right)$$
$$+ (\kappa_{a,\nu}^* + \kappa_{s,\nu})\sum_{n=1}^{\infty} I_{\nu,n} P_n + \kappa_{s,\nu} B_\nu(T) - j_{s,\nu} = 0. \tag{A1.14'}$$

Equating coefficients of the corresponding polynomials leads to the following series of equations:

$$\frac{1}{3}\frac{1}{\rho}\frac{dI_{\nu,1}}{dr} + \kappa_{s,\nu} B_\nu(T) - j_{s,\nu} = 0, \tag{A1.15}$$

assuming isotropic scattering, that is, $j_{s,\nu}$ independent of θ,

$$\frac{1}{\rho}\frac{dB_\nu(T)}{dr} + \frac{2}{5}\frac{1}{\rho}\frac{dI_{\nu,2}}{dr} + (\kappa^*_{a,\nu} + \kappa_{s,\nu})I_{\nu,1} = 0, \qquad (A1.16.1)$$

$$\frac{2}{3}\frac{1}{\rho}\frac{dI_{\nu,1}}{dr} + \frac{3}{7}\frac{1}{\rho}\frac{dI_{\nu,3}}{dr} + (\kappa^*_{a,\nu} + \kappa_{s,\nu})I_{\nu,2} = 0, \qquad (A1.16.2)$$

or, generally, for $n \geq 1$,

$$\frac{n}{2n-1}\frac{1}{\rho}\frac{dI_{\nu,n-1}}{dr} + \frac{n+1}{2n+3}\frac{1}{\rho}\frac{dI_{\nu,n+1}}{dr} + (\kappa^*_{a,\nu} + \kappa_{s,\nu})I_{\nu,n} = 0. \qquad (A1.16.n)$$

We may evaluate the ratio between successive terms of the expansion by replacing $dI_{\nu,n-1}$ by $I_{\nu,n-1}$ and dr by R (the stellar radius) in equations (A1.16.n) and neglecting factors of the order of unity. Thus to order of magnitude, we have

$$\frac{1}{\kappa\rho R}(B_\nu(T) + I_{\nu,2}) \approx I_{\nu,1} \qquad (A1.17.1)$$

$$\frac{1}{\kappa\rho R}(I_{\nu,1} + I_{\nu,3}) \approx I_{\nu,2} \qquad (A1.17.2)$$

and, generally, for $n \geq 1$,

$$\frac{1}{\kappa\rho R}(I_{\nu,n-1} + I_{\nu,n+1}) \approx I_{\nu,n}. \qquad (A1.17.n)$$

Since the deviation from isotropy is small in stellar interiors, there is some $\varepsilon < 1$, such that $I_{\nu,n} < \varepsilon B_\nu(T)$ for all $n \geq 1$. The question is how many terms of the expansion should we retain. The following argument is due to Eddington. In *all* equations (A1.17.n) with $n \geq 2$, the left-hand side is smaller than $\varepsilon B_\nu(T)/(\kappa\rho R)$, neglecting factors of the order of unity. But

$$\frac{1}{\kappa\rho R} \approx \frac{R^2}{\kappa M} \approx 10^{-10},$$

for average opacities κ and typical stellar densities and radii, and hence for $I_{\nu,2}$ and *all* subsequent coefficients we have

$$I_{\nu,n \geq 2} < 10^{-10}\varepsilon B_\nu(T).$$

We now repeat the argument using this result in relation (A1.17.n) with $n \geq 3$ and obtain $I_{\nu,n \geq 3} < 10^{-20}\varepsilon B_\nu(T)$ and again, $I_{\nu,n \geq 4} < 10^{-30}\varepsilon B_\nu(T)$, and so forth. As to $I_{\nu,1}$, from relation (A1.17.1) it follows that it is of the order of $10^{-10}B_\nu(T)$.

Clearly, the power series (A1.13) converges very rapidly,

$$\left|\frac{I_{\nu,n}}{I_{\nu,n-1}}\right| \approx 10^{-10},$$

meaning that the deviation from isotropy is indeed very small and we may discard all but the first two terms of the expansion. (Obviously, we cannot discard the second term as well, for that would leave us with an isotropic radiation field with no net flux.) This approximation is called the *diffusion approximation*. The solution of the transfer equation (A1.12) is thus

$$I_\nu(r,\theta) = B_\nu(T) + I_{\nu,1}(r)\cos\theta = B_\nu(T) - \frac{1}{(\kappa^*_{a,\nu} + \kappa_{s,\nu})\rho}\frac{dB_\nu(T)}{dr}\cos\theta, \tag{A1.18}$$

where we have eliminated $I_{\nu,1}$ from equation (A1.16.1). Finally,

$$\frac{dB_\nu(T)}{dr} = \frac{dB_\nu}{dT}\frac{dT}{dr}. \tag{A1.19}$$

For the theory of stellar structure, knowing $I_\nu(r,\theta)$ does not suffice; we are interested in $H(r)$ (introduced in Section 3.7) – the total radiation flux (in all frequencies) in the radial direction. In order to eliminate the dependence on θ, we consider *moments* of the radiation intensity field $I(r,\theta)$, which relate to the physical quantities that we have already encountered. The flux $H(r)$ is obviously given by

$$H(r) = \int I(r,\theta)\cos\theta\,d\omega. \tag{A1.20}$$

Inserting equation (A1.18) in definition (A1.1) and noting that $\int\cos\theta\,d\omega = 0$, we have

$$H(r) = -\frac{4\pi}{3\rho}\frac{dT}{dr}\int_0^\infty \frac{1}{\kappa^*_{a,\nu} + \kappa_{s,\nu}}\frac{dB_\nu(T)}{dT}d\nu. \tag{A1.20'}$$

The radiation pressure P_{rad} (introduced in Section 3.4) is due to the fact that each photon carries a momentum $h\nu/c$. Hence the radiation flux in the θ direction across a surface element dS transfers momentum of amount $I(r,\theta)\cos\theta/c$ in the radial direction, incident on an area element $dS\cos\theta$ perpendicular to it. The resulting pressure in the radial direction is therefore given by the next moment of $I(r,\theta)$:

$$P_{\text{rad}}(r) = \frac{1}{c}\int I(r,\theta)\cos^2\theta\,d\omega, \tag{A1.21}$$

leading, with equation (A1.18) and $\int \cos^3 \theta d\omega = 0$, to

$$P_{\text{rad}}(r) = \frac{1}{c} \int_0^\infty \frac{4\pi}{3} B_\nu(T) d\nu = \tfrac{1}{3} a T^4. \qquad (A1.21')$$

Finally, differentiating P_{rad} with respect to r,

$$\frac{dP_{\text{rad}}}{dr} = \frac{4\pi}{3c} \frac{dT}{dr} \int_0^\infty \frac{dB_\nu(T)}{dT} d\nu, \qquad (A1.22)$$

and dividing equation (A1.20′) by equation (A1.21′), we obtain

$$H = -\frac{c}{\bar{\kappa}\rho} \frac{dP_{\text{rad}}}{dr}, \qquad (A1.23)$$

where

$$\frac{1}{\bar{\kappa}} = \frac{\int_0^\infty \frac{1}{\kappa_{a,\nu}(1-e^{-h\nu/kT})+\kappa_{s,\nu}} \frac{dB_\nu(T)}{dT} d\nu}{\int_0^\infty \frac{dB_\nu(T)}{dT} d\nu} \qquad (A1.24)$$

is called the *Rosseland mean opacity*, after its originator, Svein Rosseland. Substituting P_{rad} from equation (A1.21′), we finally obtain the diffusion equation for radiation in the simple form:

$$H = -\frac{4ac\, T^3}{3\bar{\kappa}\rho} \frac{dT}{dr}. \qquad (A1.25)$$

It is the same as equation (3.67), derived from simplistic arguments, but it includes a rigorous treatment of the interaction between matter and radiation, expressed by $\bar{\kappa}$, which is the essence of the behaviour of stellar matter. We note that the harmonic nature of the Rosseland mean gives highest weight to the lowest opacities. At the same time, the weighting factor dB_ν/dT becomes small at very low and very high frequencies; it peaks at $\nu = 4kT/h$. In the Sun, for example, the corresponding wavelength $\lambda = c/\nu$ is about 6000 Å (within the visible range) at the surface, where $T \approx 6000$ K, and about 2.4 Å (in the X-ray range) at the centre, where $T \approx 1.5 \times 10^7$ K. The optimal radiative transfer efficiency would be attained, if the lowest opacities occurred at frequencies near $4kT/h$. This, however, is not necessarily the case.

Appendix 2

Solutions to all the exercises

Exercise 1.1: Let M be the total mass of the system, $M = M_1 + M_2$, and a the separation. Denoting by a_1 and a_2 the distances of the two stars from the centre of mass, respectively, we have $a = a_1 + a_2$, $a_1/a = M_2/M$ and $a_2/a = M_1/M$. Let ω be the common angular velocity of the stars (which is constant, since the orbits are circular), $\omega = 2\pi/P_{\text{orb}}$. Denoting by v_1 and v_2 the velocities of the two stars, respectively, we have $v_{o,1} = v_1 \sin i$ and $v_{o,2} = v_2 \sin i$. In addition,

$$\frac{2\pi a_1}{v_1} = \frac{2\pi}{\omega} \quad \text{and} \quad \frac{2\pi a_2}{v_2} = \frac{2\pi}{\omega}$$

and hence

$$\frac{M_2}{M_1} = \frac{a_1}{a_2} = \frac{v_1}{v_2} = \frac{v_{o,1}}{v_{o,2}},$$

which provides one relation between the desired masses and observables. Another relation is obtained from the equations of motion of the two stars,

$$M_1 \omega^2 a_1 = \frac{G M_1 M_2}{a^2} \quad \text{and} \quad M_2 \omega^2 a_2 = \frac{G M_2 M_1}{a^2},$$

which we may add to obtain Kepler's third law:

$$\omega^2 = \frac{GM}{a^3}.$$

Substituting $a = a_1 + a_2 = v_1/\omega + v_2/\omega = (v_{o,1} + v_{o,2})/(\omega \sin i)$, we have

$$M_1 + M_2 = \frac{P_{\text{orb}}}{2\pi G} \frac{(v_{o,1} + v_{o,2})^3}{\sin^3 i}.$$

Exercise 1.2: Consider a mass element Δm containing 10 000 hydrogen atoms and let the mass unit be the mass of a hydrogen atom. Then

$$\Delta m \approx 10\,000 \times 1 + 1000 \times 4 + 8 \times 16 + 4 \times 12 + 1 \times 14 + 1 \times 20,$$

according to the data given in the text (since elements heavier than neon are neglected, a small error is introduced). Now, by definition,

$$X = \frac{10\,000 \times 1}{\Delta m} = 0.7037$$

$$Y = \frac{1000 \times 4}{\Delta m} = 0.2815,$$

and similarly, $Z_C = 0.0034$, $Z_N = 0.0010$, $Z_O = 0.0090$, and $Z_{Ne} = 0.0014$.

Exercise 1.3:

(a) Substituting $\rho(r)$ in equation (1.5) we have

$$m(r) = \int_0^r 4\pi r^2 \rho(r)\,dr = 4\pi \rho_c \left[\int_0^r r^2\,dr - \frac{1}{R^2} \int_0^r r^4\,dr \right]$$

$$= 4\pi \rho_c \left(\frac{r^3}{3} - \frac{r^5}{5R^2} \right).$$

(b) $M = m(R) = 8\pi \rho_c R^3 / 15$.

(c) By definition, $\bar{\rho} = \frac{M}{(4\pi/3)R^3}$ and substituting (b) for M, we obtain $\bar{\rho} = 0.4\rho_c$.

Exercise 2.1:

(a) For a uniform density, $\rho = \bar{\rho}$ and

$$m(r) = \frac{4\pi r^3}{3} \bar{\rho} = M\left(\frac{r}{R}\right)^3.$$

Substituting $m(r)$ in the hydrostatic equation (2.14) and integrating from the centre ($P = P_c$) to the surface ($P = 0$), we have

$$P_c = G\bar{\rho} \int_0^R \frac{m(r)}{r^2}\,dr = \frac{3GM^2}{8\pi R^4} > \frac{GM^2}{8\pi R^4}.$$

(b) Using $\rho(r)$, $m(r)$ and $M(R)$ from Exercise 1.3, we integrate equation (2.14) to obtain

$$P_c = G \int_0^R \rho \frac{m}{r^2}\,dr = 4\pi G\rho_c^2 \int_0^R \left[1 - \left(\frac{r}{R}\right)^2\right]\left(\frac{r}{3} - \frac{r^3}{5R^2}\right)dr$$

$$= \frac{15GM^2}{16\pi R^4} > \frac{GM^2}{8\pi R^4}.$$

Exercise 2.2: If we imagine the star compressed into a sphere of uniform density ρ_c, the new central pressure P_c' must exceed P_c, since by bringing the matter closer together we increase the gravitational attraction between its parts, that is, the force to be balanced by this pressure. The new central pressure is obtained, as in *Exercise 2.1*, by integrating the hydrostatic equation (2.14), with $m = 4\pi r^3 \rho_c/3$, up to $R = (3M/4\pi \rho_c)^{1/3}$, which yields

$$P'_c = \tfrac{1}{2}(4\pi/3)^{1/3} GM^{2/3} \rho_c^{4/3}.$$

In conclusion, $P_c < P'_c$ leads to

$$P_c < (4\pi)^{1/3} 0.347 GM^{2/3} \rho_c^{4/3}.$$

Exercise 2.3:

(a) Inserting $m(r) = 4\pi r^3 \bar\rho/3$ and $dm = 4\pi r^2 \bar\rho dr$ into equation (2.20) and performing the integration, we obtain, after eliminating $\bar\rho$,

$$\Omega = -\frac{3}{5}\frac{GM^2}{R},$$

whence $\alpha = 0.6$

(b) Using $m(r)$ from *Exercise 1.3* and $dm = \rho(r) 4\pi r^2 dr$ in equation (2.20), we obtain

$$\Omega = -4\pi G\rho_c^2 \int_0^R \left(\frac{r^3}{3} - \frac{r^5}{5R^2}\right)\left[1 - \left(\frac{r}{R}\right)^2\right] 4\pi r\, dr = -\frac{5}{7}\frac{GM^2}{R},$$

whence $\alpha = 0.71$.

Exercise 2.4: The rate of change of the energy, as given by equation (2.43), is $\dot E = -L$. Assuming hydrostatic equilibrium, we have from the virial theorem $E = \tfrac{1}{2}\Omega$ [equation (2.44)] with $\Omega = -\alpha GM^2/R$ [equation (2.27)]. Hence

$$\dot E = -\tfrac{1}{2}\alpha GM^2 \left(\frac{1}{R}\right)^{\cdot} = -L.$$

Setting $t = 0$ and $R = R_0$ at the beginning of contraction, we obtain by integration

$$\frac{1}{R} - \frac{1}{R_0} = \frac{2L}{\alpha GM^2} t,$$

which yields

$$\dot R = -\frac{R_0/\tau}{(t/\tau + 1)^2}, \qquad \tau = \frac{\alpha GM^2}{2R_0 L}.$$

For $t \gg \tau$, $\dot R \to -R_0 \tau/t^2$.

Exercise 3.1: For a degenerate electron gas to be considered perfect, the Coulomb energy per particle, ϵ_C, must be smaller than the kinetic energy, in this case, $p_0^2/2m_e$, where p_0 is given by equation (3.32). The average distance between electrons is $n_e^{-1/3}$, where n_e is the electron number density. Hence $\epsilon_C \approx e^2 n_e^{1/3}/4\pi\varepsilon_0$ and the condition is

$$\frac{e^2 n_e^{1/3}}{4\pi\varepsilon_0} < \frac{h^2}{2m_e}\left(\frac{3n_e}{8\pi}\right)^{2/3}.$$

Thus the electron number density must satisfy

$$n_e > \frac{8}{9\pi}\left(\frac{m_e e^2}{\varepsilon_0 h^2}\right)^3 \approx 6 \times 10^{28} \text{ m}^{-3}.$$

Exercise 3.2: By definition [equations (3.11) and (3.12)], $P_{\text{gas}} = \beta P$ and $P_{\text{rad}} = (1-\beta)P$, and β is assumed constant throughout the star. The specific energy of a (nonrelativistic) gas, whether ideal or degenerate, is given by equation (3.44),

$$u_{\text{gas}} = \frac{3}{2}\frac{P_{\text{gas}}}{\rho} = \frac{3}{2}\beta\frac{P}{\rho},$$

and the specific energy of radiation is given by equation (3.47),

$$u_{\text{rad}} = 3\frac{P_{\text{rad}}}{\rho} = 3(1-\beta)\frac{P}{\rho}.$$

Hence

$$u_{\text{gas}} + u_{\text{rad}} = \frac{3}{2}(2-\beta)\frac{P}{\rho} \quad \Longrightarrow \quad \frac{P}{\rho} = \frac{2}{3(2-\beta)}(u_{\text{gas}} + u_{\text{rad}}).$$

Using the virial theorem in the form (2.23), we have

$$\Omega = -3\int\frac{P}{\rho}dm = -\frac{2}{(2-\beta)}\int(u_{\text{gas}} + u_{\text{rad}})dm = -\frac{2}{(2-\beta)}U.$$

Now, $E = U + \Omega$ and, substituting the relation between Ω and U, we finally obtain

$$E = \frac{\beta}{2}\Omega = -\frac{\beta}{2-\beta}U,$$

which tends to zero when the radiation pressure predominates ($\beta \to 0$) and to the well-known relations $E = \Omega/2 = -U$, when radiation pressure is negligible ($\beta \to 1$).

Exercise 3.3: The hydrostatic equation (2.14) may be written in the form

$$\frac{dP}{dr} = -\rho g,$$

where we have used the definition of the local gravitational acceleration, $g = Gm/r^2$. Dividing both sides by $\kappa\rho$ and using the definition of optical depth $d\tau = -\kappa\rho dr$, we obtain the desired equation.

Exercise 4.1: Consider a mass element Δm of helium, half of which turns into carbon and half into oxygen, by nuclear processes that can be expressed as $3\alpha \to {}^{12}C$ and $4\alpha \to {}^{16}O$. The energy released in the first process is $Q_{3\alpha} = 7.275$ MeV (see text), while the energy released in the second is given by adding to it the energy released by α capture on a ${}^{12}C$ nucleus, 7.162 MeV (see text), amounting to $Q_{4\alpha} = 14.437$ MeV. The number of ${}^{12}C$ nuclei produced is given by

$$n({}^{12}C) = \frac{0.5\Delta m}{12 m_H}$$

and, similarly,

$$n({}^{16}O) = \frac{0.5\Delta m}{16 m_H}.$$

Hence the total energy released per unit mass is

$$Q = \frac{n({}^{12}C)Q_{3\alpha} + n({}^{16}O)Q_{4\alpha}}{\Delta m} = \frac{Q_{3\alpha}/24 + Q_{4\alpha}/32}{m_H} = 7.3 \times 10^{13} \text{ J kg}^{-1}.$$

Exercise 4.2: Using the results of *Exercises 1.3* and *2.1*, in which the same density distribution is assumed, we have

$$\rho_c = \frac{15M}{8\pi R^3} \quad \text{and} \quad P_c = \frac{15GM^2}{16\pi R^4}.$$

Combining these results, and using the equation of state for an ideal gas (3.28), we obtain the central temperature

$$T_c = \frac{1}{2}\frac{\mu G}{\mathcal{R}}\frac{M}{R}, \quad \text{(Ex. 4.2.1)}$$

where $\mu = 0.61$ for a solar composition (see Section 3.3). The assumption of nondegeneracy implies that for the electrons, the ideal gas pressure (3.27) is higher than the degeneracy pressure (3.34),

$$\frac{\mathcal{R}}{\mu_e}\rho_c T_c > K_1'\left(\frac{\rho_c}{\mu_e}\right)^{5/3}, \quad \text{(Ex. 4.2.2)}$$

where $\mu_e \approx 1.17$ for a solar composition (see Section 3.3). Using (Ex. 4.2.1), we express

ρ_c in terms of T_c and M,

$$\rho_c = \frac{15}{\pi} \left(\frac{\mathcal{R}}{\mu G}\right)^3 \frac{T_c^3}{M^2},$$

and insert the expression into inequality (Ex. 4.2.2). We thus obtain an upper limit for T_c, given the stellar mass M:

$$T_c < \left(\frac{\pi}{15}\right)^{2/3} \frac{\mu^2 \mu_e^{2/3} G^2}{\mathcal{R} K_1'} M^{4/3}.$$

The desired lower limit for the stellar mass required for each nuclear burning process is obtained by reversing this relation and substituting for T_c the appropriate threshold temperatures given in Table 4.1.

Exercise 5.1: If we adopt r as the independent space variable, the Taylor expansion near $r = 0$ for any function $f(r)$ is

$$f = f_c + \left(\frac{df}{dr}\right)_c r + \frac{1}{2}\left(\frac{d^2 f}{dr^2}\right)_c r^2 + \frac{1}{6}\left(\frac{d^3 f}{dr^3}\right)_c r^3 + \cdots$$

and we retain only the first nonvanishing term besides f_c. For the mass $m(r)$ we have $m_c = 0$ (boundary condition) and from equation (5.2) on the left,

$$\left(\frac{dm}{dr}\right)_c = 4\pi (r^2 \rho)_c = 0$$

$$\left(\frac{d^2 m}{dr^2}\right)_c = 4\pi \left(2r\rho + r^2 \frac{d\rho}{dr}\right)_c = 0$$

$$\left(\frac{d^3 m}{dr^3}\right)_c = 4\pi \left(2\rho + 4r\frac{d\rho}{dr} + r^2 \frac{d^2\rho}{dr^2}\right)_c = 8\pi \rho_c.$$

Therefore near the centre

$$m(r) = \tfrac{4}{3}\pi \rho_c r^3,$$

as if the density were uniform and equal to the central value. For the pressure $P(r)$ we have from equation (5.1) on the left and the result obtained for $m(r)$

$$\left(\frac{dP}{dr}\right)_c = -\left(\rho \frac{Gm}{r^2}\right)_c = -\left(\frac{4\pi G \rho^2 r}{3}\right)_c = 0$$

$$\left(\frac{d^2 P}{dr^2}\right)_c = -\left(\frac{d\rho}{dr}\frac{Gm}{r^2} + \rho \frac{G}{r^2}\frac{dm}{dr} - 2\rho \frac{Gm}{r^3}\right)_c = -\frac{4\pi G \rho_c^2}{3}.$$

Therefore near the centre

$$P(r) = P_c - \tfrac{2}{3}\pi G \rho_c^2 r^2.$$

For the luminosity $F(r)$ we have $F_c = 0$ (boundary condition) and from equation (5.4) on the left

$$\left(\frac{dF}{dr}\right)_c = 4\pi (r^2 \rho q)_c = 0$$

$$\left(\frac{d^2 F}{dr^2}\right)_c = 4\pi \left(2r\rho q + r^2 \frac{d\rho}{dr} q + r^2 \rho \frac{dq}{dr}\right)_c = 0$$

$$\left(\frac{d^3 F}{dr^3}\right)_c = 4\pi [2\rho q + r(\ldots) + r^2(\ldots)]_c = 8\pi \rho_c q_c.$$

Therefore near the centre

$$F(r) = \tfrac{4}{3}\pi \rho_c q_c r^3.$$

For the temperature $T(r)$ we have from equation (5.3) on the left and the result obtained for $F(r)$

$$\left(\frac{dT}{dr}\right)_c = -\frac{3}{16\pi ac}\left(\frac{\kappa \rho}{T^3}\frac{F}{r^2}\right)_c = -\frac{1}{4ac}\left(\frac{\kappa \rho^2 q}{T^3} r\right)_c = 0$$

$$\left(\frac{d^2 T}{dr^2}\right)_c = -\frac{3}{16\pi ac}\left[\frac{\kappa \rho}{T^3}\frac{d}{dr}\left(\frac{F}{r^2}\right) + \frac{F}{r^2}\frac{d}{dr}\left(\frac{\kappa \rho}{T^3}\right)\right]_c = -\frac{1}{4ac}\frac{\kappa_c \rho_c^2 q_c}{T_c^3}.$$

Therefore near the centre

$$T(r) = T_c - \frac{1}{8ac}\frac{\kappa_c \rho_c^2 q_c}{T_c^3} r^2.$$

Note that these relations hold regardless of the functional dependences $P(\rho, T)$, $q(\rho, T)$, and $\kappa(\rho, T)$.

Exercise 5.2:

(a) For $n = 0$, the Lane–Emden equation (5.17) becomes

$$\frac{d}{d\xi}\left(\xi^2 \frac{d\theta}{d\xi}\right) = -\xi^2.$$

Integrating, we obtain

$$\xi^2 \frac{d\theta}{d\xi} = -\frac{1}{3}\xi^3 + C,$$

where C is an integration constant. Dividing by ξ^2 and integrating again, we obtain the solution

$$\theta = -\frac{1}{6}\xi^2 - \frac{C}{\xi} + D,$$

where D is a second integration constant. Since we cannot accept solutions that are singular at the origin, we must assume $C = 0$, and since $\theta = 1$ at the origin (by definition), $D = 1$. The solution for $n = 0$ is therefore

$$\theta(\xi) = 1 - \tfrac{1}{6}\xi^2.$$

Obviously, $\xi_1 \equiv \xi(\theta = 0) = \sqrt{6}$ and $(d\theta/d\xi)_{\xi_1} = -\xi_1/3 = -\sqrt{2/3}$. Substituting into equation (5.20) and using equation (5.18) to eliminate α, we obtain $M = 4\pi R^3 \rho_c/3$ (and $D_0 = 1$), which shows that an $n = 0$ polytrope describes a configuration of uniform density.

(b) For $n = 1$ and a variable χ defined as $\chi = \xi\theta$, the Lane–Emden equation (5.17) becomes

$$\frac{d^2\chi}{d\xi^2} = -\chi,$$

whose general solution is

$$\chi = C\sin(\xi - \delta),$$

where C and δ are constants of integration. Hence

$$\theta = \frac{C\sin(\xi - \delta)}{\xi}.$$

We must assume $\delta = 0$, for otherwise the solution is singular at the origin, and since $\theta = 1$ at the origin, $C = 1$. The solution for $n = 1$ is therefore

$$\theta(\xi) = \frac{\sin\xi}{\xi},$$

which has its first zero at $\xi_1 = \pi$ [and is monotonically decreasing in the interval $(0, \pi)$]. Differentiating, we obtain

$$\left(\frac{d\theta}{d\xi}\right)_{\xi_1} = \left(\frac{\cos\xi}{\xi} - \frac{\sin\xi}{\xi^2}\right)_{\xi=\pi} = -\frac{1}{\pi}.$$

We now use equations (5.18) and (5.20) to obtain $M = 4R^3 \rho_c/\pi$ (noting that $D_1 = \pi^2/3$, consistent with the entry in Table 5.1).

Exercise 5.3: For given M and P_c, we have from equation (5.28)

$$\frac{\rho_{c,1.5}}{\rho_{c,3}} = \left(\frac{B_3}{B_{1.5}}\right)^{3/4}.$$

For given M, we obtain the ratio of radii $R(n)$ from equation (5.21) and Table 5.1:

$$\frac{R(1.5)}{R(3)} = \left(\frac{D_{1.5}}{D_3}\frac{\rho_{c,3}}{\rho_{c,1.5}}\right)^{1/3} = \left(\frac{D_{1.5}}{D_3}\right)^{1/3}\left(\frac{B_{1.5}}{B_3}\right)^{1/4} = \left(\frac{5.991}{54.81}\right)^{1/3}\left(\frac{0.206}{0.157}\right)^{1/4} < 1$$

and therefore

$$R(3) > R(1.5).$$

Exercise 5.4: The central density is readily given by equation (5.21): $\rho_c = 1.2 \times 10^2 \text{ kg m}^{-3}$. In order to obtain the central pressure as a function of M and R, we eliminate ρ_c between equations (5.21) and (5.28):

$$P_c = \frac{GM^2}{4\pi R^4}\left[(3D_n)^{4/3} B_n\right].$$

The term in square brackets exceeds unity for all n and hence

$$P_c > \frac{GM^2}{4\pi R^4} > \frac{GM^2}{8\pi R^4}.$$

Thus inequality (2.18) is generally satisfied by polytropic models. For *Capella*, with $n = 3$, $P_c = 6.1 \times 10^{12} \text{ N m}^{-2}$.

Exercise 5.5: The critical mass is obtained from the relativistic-degenerate equation of state (3.36). Hence at the stellar centre both equations (5.28) and (3.36) are satisfied, both being of the form $P_c \propto \rho_c^{4/3}$. Equating coefficients and isolating M, we obtain

$$M = \left[B_3^{-3/2}\frac{1}{8^{3/2}}\left(\frac{3}{\pi}\right)^{1/2}\frac{1}{(4\pi)^{1/2}}\right]\left(\frac{hc}{Gm_H^{4/3}}\right)^{3/2}\mu_e^{-2}.$$

The term in square brackets reduces to $B_3^{-3/2}\sqrt{1.5}/32\pi$.

Exercise 5.6: In radiative equilibrium, the radiation pressure gradient is obtained from equations (5.3) and (3.40):

$$\frac{dP_{\rm rad}}{dr} = -\frac{\kappa\rho}{c}\frac{F}{4\pi r^2}.$$

(In the case of convection, this relation is still correct, provided the flux F on the right-hand side is taken to be the radiative flux, rather than the total flux, of which the bulk is due to convection.) Substituting into the hydrostatic equation (5.1) $P = P_{\rm gas} + P_{\rm rad}$, we obtain

$$\frac{dP_{\rm gas}}{dr} = -\rho\frac{GM}{r^2} + \frac{\kappa\rho F}{4\pi r^2 c} = -\rho\frac{GM}{r^2}\left(1 - \frac{\kappa F}{4\pi c Gm}\right).$$

So long as condition (5.34) is satisfied, the gas pressure decreases outward. When it is violated, the density is bound to increase outward, if the temperature is decreasing outward. This would lead to instability (of the Rayleigh–Taylor type).

Exercise 5.7:

(a) Equation (5.24) for an $n = 3$ polytrope may be written as

$$M^2 = \frac{(4\pi M_3)^2}{(\pi G)^3}K^3,$$

where M_3 is given in Table 5.1. Substituting K from equation (5.45), we obtain the quartic equation in the form

$$1 - \beta = \left(\frac{M}{M_\star}\right)^2 \mu^4\beta^4,$$

where

$$M_\star = \frac{4M_3\mathcal{R}^2}{\sqrt{\pi a/3}\, G^{3/2}}.$$

With $a = 8\pi^5 k^4/(15c^3 h^3)$ and $\mathcal{R} = k/m_{\rm H}$, we have

$$M_\star = \frac{3\sqrt{10}M_3}{\pi^3}\left(\frac{hc}{Gm_{\rm H}^{4/3}}\right)^{3/2} = 18.3 M_\odot.$$

(b) The Chandrasekhar mass, given by equation (5.31), may be expressed as

$$M_{\rm Ch} = \frac{\pi^2}{8\sqrt{15}}\mu_e^{-2} M_\star.$$

Exercise 6.1:

(a) Assume a white dwarf of mass M and radius $R(M)$ has an outer layer of solar composition and of mass $\Delta m \ll M$ (and negligible thickness). The energy required to expel this layer is equal to the gravitational binding energy $GM\Delta m/R(M)$. If Q is the energy released per unit mass of burnt hydrogen (from Section 4.3, $Q \approx 6 \times 10^{14}$ J kg^{-1}), and the hydrogen mass fraction in the outer layer is $X_\odot \approx 0.7$, then the amount of hydrogen mass burnt is $f \Delta m X_\odot$, satisfying

$$\frac{GM\Delta m}{R(M)} = f \Delta m X_\odot Q.$$

(b) The $R(M)$ relationship for white dwarfs (5.29), appropriate to a nonrelativistic equation of state, that is, for $M < M_{\text{Ch}}$, may be calibrated with the aid of the provided data:

$$\frac{R}{0.01 R_\odot} = \left(\frac{M}{M_\odot}\right)^{-1/3}.$$

Combining these results, we have

$$f = \frac{GM_\odot}{0.01 R_\odot X_\odot Q} \left(\frac{M}{M_\odot}\right)^{4/3} \approx 0.045 \left(\frac{M}{M_\odot}\right)^{4/3}.$$

Note that for typical white dwarf masses this fraction is very small, despite the strong gravitational field that must be overcome.

Exercise 6.2: Adiabatic processes satisfy equation (3.48):

$$du + P d\left(\frac{1}{\rho}\right) = 0 \quad \Longrightarrow \quad du = \frac{P}{\rho}\frac{d\rho}{\rho},$$

from which relations may be derived between any two of the thermodynamic functions P, ρ, and T. For gas and radiation we define adiabatic exponents Γ_1 and Γ_2 by

$$\frac{dP}{P} = \Gamma_1 \frac{d\rho}{\rho} \qquad \text{(Ex. 6.2.1)}$$

$$\frac{dP}{P} = \frac{\Gamma_2}{\Gamma_2 - 1} \frac{dT}{T}, \qquad \text{(Ex. 6.2.2)}$$

noting that both are equal to the γ_a of conditions (6.26) and (6.28) in the case of gas without radiation. Now, for an ideal gas we have from equations (3.28), (3.44), and (3.47)

$$u = \frac{3}{2}\frac{\mathcal{R}}{\mu}T + \frac{aT^4}{\rho}$$

$$P = P_{\text{gas}} + P_{\text{rad}} = \frac{\mathcal{R}}{\mu}\rho T + \tfrac{1}{3}aT^4$$

and from equations (3.11) and (3.12),

$$P_{\text{gas}} = \beta P \quad \text{and} \quad P_{\text{rad}} = (1-\beta)P.$$

Hence

$$du = \frac{3}{2}\frac{\mathcal{R}}{\mu}dT + \frac{4aT^3}{\rho}dT - \frac{aT^4}{\rho^2}d\rho = \frac{3}{2}\beta\frac{P}{\rho}\frac{dT}{T} + 12(1-\beta)\frac{P}{\rho}\frac{dT}{T} - 3(1-\beta)\frac{P}{\rho}\frac{d\rho}{\rho},$$

which, substituted into the condition for adiabaticity, leads to

$$\frac{24 - 21\beta}{2}\frac{dT}{T} = (4 - 3\beta)\frac{d\rho}{\rho}. \qquad \text{(Ex. 6.2.3)}$$

For the pressure we have

$$dP = P_{\text{gas}}\frac{dT}{T} + P_{\text{gas}}\frac{d\rho}{\rho} + 4P_{\text{rad}}\frac{dT}{T},$$

leading to

$$\frac{dP}{P} = (4 - 3\beta)\frac{dT}{T} + \beta\frac{d\rho}{\rho}. \qquad \text{(Ex. 6.2.4)}$$

Eliminating dT/T between equations (Ex. 6.2.3) and (Ex. 6.2.4), we obtain

$$\frac{dP}{P} = \left[\frac{2(4 - 3\beta)^2}{24 - 21\beta} + \beta\right]\frac{d\rho}{\rho}.$$

Comparing this result with equation (Ex. 6.2.1), we have

$$\Gamma_1 = \frac{32 - 24\beta - 3\beta^2}{24 - 21\beta}.$$

Similarly, by eliminating $d\rho/\rho$ between equations (Ex. 6.2.3) and (Ex. 6.2.4) and comparing with equation (Ex. 6.2.2), we obtain

$$\Gamma_2 = \frac{32 - 24\beta - 3\beta^2}{24 - 18\beta - 3\beta^2}.$$

For $\beta = 1$ (pure gas), $\Gamma_1 = \Gamma_2 = 5/3$; for $\beta = 0$ (pure radiation), $\Gamma_1 = \Gamma_2 = 4/3$; for $\beta = \frac{1}{2}$, $\Gamma_1 = 1.43$, while $\Gamma_2 = 1.35$.

Note: The adiabatic exponent Γ_1 for matter and radiation was introduced by Eddington in 1918; Γ_2, as well as a further adiabatic exponent Γ_3, which relates T and ρ, were later introduced by Chandrasekhar.

Exercise 6.3: Let M_c be the mass of the convective core. The temperature gradient at its boundary is given on the one hand by the adiabatic gradient (as in the core),

$$\frac{dT}{dr} = \frac{\gamma_a - 1}{\gamma_a} \frac{T}{P} \frac{dP}{dr} = -\frac{\gamma_a - 1}{\gamma_a} \frac{T}{P} \frac{GM_c \rho}{r^2},$$

after substituting the pressure gradient from the hydrostatic equation, and on the other hand by the radiative diffusion equation (5.3),

$$\frac{dT}{dr} = -\frac{3}{4ac} \frac{\kappa \rho}{T^3} \frac{F}{4\pi r^2}.$$

Continuity of dT/dr (imposed by the continuity of the radiative flux) requires equality of the right-hand sides of these equations:

$$\frac{\gamma_a - 1}{\gamma_a} \frac{T}{P} GM_c = \frac{3}{4ac} \frac{\kappa}{T^3} \frac{F}{4\pi}.$$

Since there are no energy sources outside the core, we may take $F = L$. Substituting

$$\frac{\frac{1}{3}aT^4}{P} = \frac{P_{\rm rad}}{P} = 1 - \beta,$$

dividing by M, and rearranging terms, we obtain

$$\frac{M_c}{M} = \frac{\gamma_a}{4(\gamma_a - 1)(1 - \beta)} \frac{\kappa L}{4\pi c GM}.$$

Now, if κ is constant up to the surface, then $4\pi cGM/\kappa$ is the Eddington luminosity $L_{\rm Edd}$, and if β is constant, we have from equation (5.42) $L/L_{\rm Edd} = 1 - \beta$, which yields the desired expression for the core mass fraction. Note that the ratio depends indirectly on M through the adiabatic exponent, which depends on β, where $\beta = \beta(M)$.

Exercise 7.1: Inserting relation (7.36) into equation (7.28), on the right we obtain

$$P_\star \propto M^2 \left(M^{\frac{n-1}{n+3}}\right)^{-4} \quad \Longrightarrow \quad P_\star \propto M^{\frac{10-2n}{n+3}}.$$

We have $P_\star \propto M^{2/7}$ for $n = 4$ (that is, $P_\star$ increases with M), whereas $P_\star \propto M^{-22/19}$ for $n = 16$ (that is, $P_\star$ decreases with increasing stellar mass).

Inserting relation (7.36) into equation (7.33), we obtain

$$T_\star \propto M^{\frac{4}{n+3}},$$

which yields $T_\star \propto M^{4/7}$ for $n = 4$ and $T_\star \propto M^{4/19}$ for $n = 16$. Note the weak dependence of $T_\star$ on M corresponding to stars that burn hydrogen by the CNO cycle, which means that the main sequence of these stars may be taken to represent a line of constant central temperature.

Exercise 7.2: The effective temperature of a star of known L and R is obtained from equation (1.3):

$$L = 4\pi R^2 \sigma T_{\text{eff}}^4.$$

Using relation (7.35) for the luminosity at the lower end of the main sequence,

$$\frac{L_{\min}}{L_\odot} = \left(\frac{M_{\min}}{M_\odot}\right)^3,$$

and relation (7.36) for the radius (calibrated to the solar radius, with $n = 4$),

$$\frac{R}{R_\odot} = \left(\frac{M_{\min}}{M_\odot}\right)^{3/7},$$

we obtain

$$L_\odot \left(\frac{M_{\min}}{M_\odot}\right)^3 = 4\pi R_\odot^2 \left(\frac{M_{\min}}{M_\odot}\right)^{6/7} \sigma T_{\text{eff,min}}^4.$$

Substituting $T_{\text{eff},\odot} = [L_\odot/(4\pi R_\odot^2 \sigma)]^{1/4} \approx 5800$ K, and $M_{\min} \approx 0.1 M_\odot$, we have

$$T_{\text{eff,min}} = T_{\text{eff},\odot} \left(\frac{M_{\min}}{M_\odot}\right)^{15/28} \approx 1700 \text{ K}.$$

Exercise 7.3: First, we write the condition $L < 4\pi c G M / \kappa_s$ as

$$\frac{L}{L_\odot} < \frac{4\pi c G M_\odot}{\kappa_s L_\odot} \frac{M}{M_\odot}.$$

Next, we calibrate relation (7.35) between luminosity and mass:

$$\frac{L}{L_\odot} = \left(\frac{M}{M_\odot}\right)^3.$$

Substituting it into the previous relation, we obtain an upper limit for the mass of main-sequence stars,

$$\frac{M}{M_\odot} < \sqrt{\frac{4\pi c G M_\odot}{\kappa_s L_\odot}} = 180,$$

assuming κ_s is the electron scattering opacity $\kappa_{es,o}$ [equation (3.64)]. Using the mass-luminosity relation, we obtain the corresponding upper limit for the luminosity of main-sequence stars: $L < 5.8 \times 10^6 \, L_\odot$. The radius of a $180 M_\odot$ star may be obtained from calibrated relation (7.36), taking $n = 16$, appropriate to the upper main sequence. The effective temperature results from $L = 4\pi R^2 \sigma T_{eff}^4$, as in *Exercise 7.2*, which yields $T_{eff} = 3.7 \times 10^4$ K.

Exercise 7.4: Some of the relations between starred quantities [equations (7.28), (7.29), and (7.33)] are independent of the opacity or the nuclear energy generation laws. These are

$$P_\star = \frac{GM^2}{R_\star^4} \qquad \rho_\star = \frac{M}{R_\star^3} \qquad T_\star = \frac{G\mu}{\mathcal{R}} \frac{M}{R_\star}. \qquad \text{(Ex. 7.4.1)}$$

From the Kramers opacity law and $n = 4$ we obtain two additional relations, using equations (7.31) and (7.32):

$$F_\star = \frac{ac}{\kappa_0} \frac{T_\star^{7.5} R_\star^4}{\rho_\star M}$$

$$F_\star = q_0 \rho_\star T_\star^4 M.$$

Substituting (Ex. 7.4.1), we have

$$F_\star \propto \frac{M^{5.5}}{R_\star^{0.5}} \qquad \text{and} \qquad F_\star \propto \frac{M^6}{R_\star^7},$$

which, combined, yield a relation between radius and mass in the form

$$R_\star \propto M^{1/13}.$$

This, in turn, enables the derivation of a mass-luminosity relation, as well as a radius-luminosity relation. With the aid of the latter, the main-sequence slope may be derived as in the text. Thus,

$$L \propto M^{5.46}$$

$$\log L = 4.12 \log T_{eff} + \text{constant}.$$

In conclusion, different opacity laws result in different main-sequence slopes (even assuming the same n), 4.12 for a Kramers opacity law, as compared to 5.6 for a constant opacity [equation (7.39a)].

Exercise 7.5: The set of structure equations in the case of a fully convective star is

$$\frac{dP}{dm} = -\frac{Gm}{4\pi r^4}$$

$$\frac{dr}{dm} = \frac{1}{4\pi r^2 \rho}$$

$$\frac{dF}{dm} = q_0 \rho T^n$$

$$P = K_a \rho^{\gamma_a}$$

$$T = K'_a P^{(\gamma_a - 1)/\gamma_a}.$$

The first three are the same as in the case of a star in radiative equilibrium and lead to the relations

$$P_\star = \frac{GM^2}{R_\star^4} \qquad \text{(Ex. 7.5.1)}$$

$$\rho_\star = \frac{M}{R_\star^3} \qquad \text{(Ex. 7.5.2)}$$

$$F_\star = q_0 \rho_\star T_\star^n M. \qquad \text{(Ex. 7.5.3)}$$

From the last two structure equations, substituting $\gamma_a = 5/3$, we obtain the relations

$$P_\star \propto \rho_\star^{5/3} \qquad \text{(Ex. 7.5.4)}$$

$$T_\star \propto P_\star^{2/5}. \qquad \text{(Ex. 7.5.5)}$$

Substitution of equations (Ex. 7.5.1) and (Ex. 7.5.2) into relation (Ex. 7.5.4) leads to a relation between radius and mass:

$$R_\star \propto M^{-1/3} \qquad \text{(Ex. 7.5.6)}$$

[compare with relation (5.29)]. Inserting it into equations (Ex. 7.5.1) and (Ex. 7.5.2), and using relation (Ex. 7.5.5), we have

$$\rho_\star \propto M^2 \qquad P_\star \propto M^{10/3} \qquad T_\star \propto M^{4/3}.$$

Using the first and last of these relations in equation (Ex. 7.5.3), we get the mass-luminosity relation

$$F_\star \propto M^{3+4n/3} \quad \Longrightarrow \quad L \propto M^{3+4n/3}. \qquad \text{(Ex. 7.5.7)}$$

Combining relations (Ex. 7.5.6) and (Ex. 7.5.7), we have for the stellar radius

$$R \propto L^{-1/(9+4n)},$$

which leads to the main-sequence slope (since $T_{\text{eff}}^4 \propto L/R^2$):

$$\log L = \frac{36 + 16n}{11 + 4n} \log T_{\text{eff}} + \text{constant}.$$

We note that for the lower main sequence ($n \approx 4$), where we may indeed expect the stars to be fully convective (see Section 6.6), the slope is 3.7 – less steep than in the case when radiative equilibrium is assumed.

Exercise 7.6:

(a) Assume an amount of mass δm is burnt during a time interval δt (and added to the core). The nuclear energy supplied is $Q\,\delta m$; this energy is radiated by the star at a rate L and hence $Q\,\delta m = L\,\delta t$. Therefore the rate of core growth is $\dot{M}_c = L/Q$. Since L and Q are constants, and $M_c = 0$ at $t = 0$, we get by integration

$$M_c(t) = \frac{L}{Q} t. \qquad \text{(Ex. 7.6.1)}$$

(b) The envelope loses mass at its inner boundary at the same rate as the core gains mass due to nuclear burning. It also loses mass at its outer boundary – at the mass loss rate of the star. Thus

$$\dot{M}_e = -\dot{M}_c + \dot{M} = -\frac{L}{Q} - \alpha L = -L\left(\frac{1}{Q} + \alpha\right).$$

Integrating and using the initial condition $M_e = M_0$ at $t = 0$, we have

$$M_e(t) = M_0 - L\left(\frac{1}{Q} + \alpha\right) t. \qquad \text{(Ex. 7.6.2)}$$

(c) The core mass attained when the envelope mass is exhausted is obtained by setting $M_e(t) = 0$ and eliminating t between equations (Ex. 7.6.1) and (Ex. 7.6.2),

$$M_c = \frac{M_0}{1 + \alpha Q}.$$

(d) For the star to become a white dwarf this core mass must satisfy $M_c < M_{Ch}$, which imposes an upper limit on the initial mass of the star:

$$M_0 < M_{Ch}(1 + \alpha Q) \quad \Longrightarrow \quad M_0 < 9\, M_\odot.$$

Exercise 8.1: The wind emanated by the Sun crosses any spherical surface centred on the Sun (just as the radiation emitted by the Sun does); otherwise matter would accumulate at some place; hence $\dot{m} = $ constant. Conservation of mass (in spherical symmetry) requires that an amount of mass δm crossing a spherical surface of radius r during a time interval δt equal the density at r multiplied by the volume of this mass, $\delta V = 4\pi r^2 \delta r$. Since $\delta r = v \delta t$, where v is the (radial) velocity of the wind, we have

$$\delta m = 4\pi r^2 \rho v \delta t.$$

Dividing by δt, we obtain

$$\dot{m} = 4\pi r^2 \rho v.$$

As the contribution of electrons to the mass (density) is negligible, we may assume the wind density to be $\rho \approx n_p m_H$, where n_p is the proton number density. The measurements at Earth ($r = 1$ AU) thus yield $\dot{m} \approx 1.3 \times 10^9$ kg s$^{-1} \approx 2 \times 10^{-14} M_\odot$ yr^{-1}.

Exercise 8.2:

(a) Assume n helium nuclei are produced in the Sun per unit time, of which n_1 are produced by the p–p I chain, n_2 by the p–p II chain, and n_3 by the p–p III chain. Thus $n = n_1 + n_2 + n_3$ and the branching ratios are n_i/n ($1 \le i \le 3$), respectively. The neutrino fluxes intercepted at Earth, $f_{\nu,i}$ ($1 \le i \le 3$) – listed in the second column of Table 8.3 – are a fraction $\alpha = (4\pi d^2)^{-1}$ (where $d = 1$ AU) of those produced per unit time in the Sun. In the production of a helium nucleus by the p–p I chain, two p–p neutrinos are emitted; by the p–p II chain, one p–p neutrino and one ^{7}Be neutrino; and by the p–p III chain, one p–p neutrino and one ^{8}B neutrino (see Section 4.3). Therefore

$$f_{\nu,1} = \alpha(2n_1 + n_2 + n_3)$$
$$f_{\nu,2} = \alpha n_2$$
$$f_{\nu,3} = \alpha n_3$$
$$f_{\nu,1} + f_{\nu,2} + f_{\nu,3} = 2\alpha n.$$

Appendix 2 Solutions to all the exercises

Eliminating n_i from these relations we obtain the branching ratios:

$$\frac{n_1}{n} = \frac{f_{\nu,1} - f_{\nu,2} - f_{\nu,3}}{f_{\nu,1} + f_{\nu,2} + f_{\nu,3}} \approx 0.85 \quad (p\text{-}p\text{ I})$$

$$\frac{n_2}{n} = \frac{2 f_{\nu,2}}{f_{\nu,1} + f_{\nu,2} + f_{\nu,3}} \approx 0.15 \quad (p\text{-}p\text{ II})$$

$$\frac{n_3}{n} = \frac{2 f_{\nu,3}}{f_{\nu,1} + f_{\nu,2} + f_{\nu,3}} \approx 2 \times 10^{-4} \quad (p\text{-}p\text{ III}).$$

(b) The average energy carried by each neutrino type, $Q_{\nu,i}$ is listed in the last column of Table 8.3. The neutrino luminosity of the Sun is given by the total neutrino energy flux at Earth, multiplied by $4\pi d^2$:

$$L_\nu = 4\pi d^2 (f_{\nu,1} Q_{\nu,1} + f_{\nu,2} Q_{\nu,2} + f_{\nu,3} Q_{\nu,3}) = 8.9 \times 10^{24} \text{ J s}^{-1} = 0.023 L_\odot.$$

(c) If the branching ratios of the p–p chain were not known, then the neutrino energy lost for each helium nucleus produced would vary between a minimum value of 2×0.263 MeV (corresponding to the p–p I chain) and a maximum value of $(0.263 + 7.2)$ MeV (corresponding to the p–p III chain). The net energy released in the production of a helium nucleus (that would ultimately be radiated by the Sun) would range between $Q_{\max} = 26.73 - 2 \times 0.263 = 26.20$ MeV and $Q_{\min} = 26.73 - 0.263 - 7.2 = 19.27$ MeV. Since the luminosity of the Sun is known, the number of helium nuclei that should be produced per unit time in order to supply it can be calculated in each case. The number of neutrinos emitted is twice as much. Therefore

$$n_{\min} = 2 \frac{L_\odot}{Q_{\max}} = 1.84 \times 10^{38} \text{ s}^{-1}$$

$$n_{\max} = 2 \frac{L_\odot}{Q_{\min}} = 2.50 \times 10^{38} \text{ s}^{-1}.$$

Exercise 8.3: First, we integrate equation (5.2) in order to obtain the core mass M_1:

$$M_1 = \int_0^{R_1} 4\pi r^2 \left[\rho_c - (\rho_c - \rho_1) \left(\frac{r}{R_1} \right)^2 \right] dr = \frac{4\pi R_1^3}{5} (\tfrac{2}{3}\rho_c + \rho_1). \quad (\text{Ex. 8.3.1})$$

Next, we integrate equation (5.2) in order to obtain the mass outside the core:

$$M - M_1 = \int_{R_1}^{R} 4\pi r^2 \rho_1 \frac{\left(\frac{R_1}{r}\right)^3 - \left(\frac{R_1}{R}\right)^3}{1 - \left(\frac{R_1}{R}\right)^3} dr = 4\pi R_1^3 \rho_1 \left[\frac{\ln \frac{R}{R_1}}{1 - \left(\frac{R_1}{R}\right)^3} - \frac{1}{3} \right]. \quad (\text{Ex. 8.3.2})$$

Dividing equation (Ex. 8.3.2) by equation (Ex. 8.3.1) and substituting $x_1 = \rho_c/\rho_1$ and

$y_1 = M/M_1$, we have

$$\frac{\ln \frac{R}{R_1}}{1 - \left(\frac{R_1}{R}\right)^3} = \tfrac{1}{5}(y_1 - 1)\left(\tfrac{2}{3}x_1 + 1\right) + \tfrac{1}{3}.$$

Now, since $R_1 < R$, we may neglect $(R_1/R)^3$ with respect to 1 in the denominator on the left-hand side; hence exponentiating, we obtain

$$\frac{R}{R_1} \approx e^{[(y_1 - 1)(2x_1 + 3) + 5]/15},$$

which yields $R/R_1 \approx 3 \times 10^4$ for $x_1 = 10$ and $y_1 = 7.5$. Thus, if the core radius is of the order of a white dwarf's, $R_1 \sim 0.01\, R_\odot$, the resulting stellar radius is $\sim 300\, R_\odot$, illustrating the possibility of having a compact core and a very extended envelope.

Exercise 8.4:

(a) The mass loss timescale may be estimated by $M/\dot M$ [equation (2.55)]. The thermal timescale is given by equation (2.59), $\tau_{\rm th} \approx GM^2/RL$. Using equation (8.27) for $\dot M$, we obtain

$$\tau_{m-1} = \frac{M}{\dot M} = \frac{1}{\phi}\frac{c}{v_{\rm esc}}\frac{GM^2}{RL} \approx \frac{1}{\phi}\frac{c}{v_{\rm esc}}\tau_{\rm th}.$$

Generally, $v_{\rm esc} \ll c$ and certainly $\phi v_{\rm esc} \ll c$; therefore we may conclude that $\tau_{m-1} \gg \tau_{\rm th}$.

(b) The energy required for removing an amount of mass δm from the surface of a star is equal to the gravitational binding energy of this element, $\delta E_{\rm grav} = GM\delta m/R$, and if the mass is removed during a time interval δt, the *rate* of energy supply $(\delta E_{\rm grav}/\delta t)$ is

$$\dot E_{\rm grav} = \frac{GM\dot M}{R} = \phi\frac{v_{\rm esc}}{c}L,$$

where $\dot M$ was substituted from equation (8.27). As argued in (a), $\dot E_{\rm grav} \ll L$.

(c) From estimate (2.61), $\tau_{\rm nuc} \approx \epsilon Mc^2/L$, where ϵ amounts to a few times 0.001. Using the result of (a), we have

$$\frac{\tau_{m-1}}{\tau_{\rm nuc}} = \frac{1}{\phi}\frac{1}{v_{\rm esc}c}\frac{GM}{\epsilon R}.$$

Substituting on the right-hand side $GM/R = v_{\rm esc}^2/2$, we obtain

$$\frac{\tau_{m-1}}{\tau_{\rm nuc}} = \frac{1}{\phi}\frac{v_{\rm esc}}{c}\frac{1}{2\epsilon}.$$

If $v_{esc} < 0.001c$ (as is mostly the case) and if ϕ is not a too small fraction (as, indeed, observations indicate), then $\tau_{m-1} < \tau_{nuc}$.

Exercise 8.5: In Section 7.4 we have seen that, for main-sequence stars, global quantities may be expressed as power laws of the stellar mass. These may be easily reverted to power laws of the luminosity. Thus

$$L \propto M^{\alpha_1} \implies M \propto L^{1/\alpha_1},$$
$$R \propto M^{\alpha_2} \implies R \propto L^{\alpha_2/\alpha_1}.$$

A parametrization of the mass loss rate of the form (8.27) would result in

$$\dot{M} \propto \frac{LR}{GM} \propto L^{(\alpha_1 + \alpha_2 - 1)/\alpha_1}.$$

Using the results of Section 7.4, we have $\alpha_1 = 3$ [relation (7.35)] and $\alpha_2 = (n-1)/(n+3)$ [relation (7.36)], whence

$$\dot{M} \propto L^{(3n+5)/(3n+9)}.$$

We note that this is very close to a linear dependence, particularly for massive stars, which burn hydrogen by means of the CNO cycle ($n \approx 16$).

Exercise 8.6:

(a) In the outer layer of a white dwarf we have by equation (8.36) $P = P(T)$. We may thus write the equation of hydrostatic equilibrium (8.32) as

$$\frac{dP}{dT}\frac{dT}{dr} = -\rho \frac{GM}{r^2}.$$

Using the ideal gas equation of state (appropriate to this layer), we substitute $\rho = (\mu/\mathcal{R})(P/T)$ to obtain

$$\frac{d \ln P}{d \ln T}\frac{dT}{dr} = -\frac{\mu}{\mathcal{R}}\frac{GM}{r^2}. \quad \text{(Ex. 8.6.1)}$$

From equation (8.36) we have

$$\frac{d \ln P}{d \ln T} = \frac{17}{4},$$

and hence, integrating equation (Ex. 8.6.1) and using the boundary condition $T(R) = 0$, we obtain the required relation (8.42).

(b) We may write this relation for $T_c = T_b \equiv T(r_b)$ in the form

$$\frac{\mathcal{R}}{\mu}T_c = \frac{4}{17}\frac{GM}{R}\frac{R-r_b}{r_b}. \qquad \text{(Ex. 8.6.2)}$$

The left-hand side represents (roughly) the ion energy per unit mass, P_I/ρ. The term GM/R on the right-hand side is, according to the virial theorem, the total energy per unit mass (or P/ρ). Since for degenerate electrons, $P \approx P_e \gg P_I$, we must have

$$R - r_b \ll r_b < R.$$

(c) Since we have shown that $\ell \equiv R - r_b \ll R$, then $r_b \approx R$ and we may write equation (Ex. 8.6.2) as

$$T_c = \frac{4}{17}\frac{\mu}{\mathcal{R}}\frac{GM}{R^2}\ell \quad\Longrightarrow\quad T_c \propto \ell.$$

Using relation (8.39) between L and T_c, we obtain $\ell \propto L^{2/7}$, and hence

$$\frac{\ell_1}{\ell_2} = \left(\frac{L_1}{L_2}\right)^{2/7} = 100^{2/7} \approx 3.7.$$

Exercise 9.1: The equation of motion for free fall (a motion governed by the gravitational field without any, or with negligible, opposition exerted by pressure) is, according to equation (2.12),

$$\ddot{r}(m,t) = -\frac{Gm}{r(m,t)^2}.$$

Multiplying both sides by $\dot{r}(m,t)$, we obtain

$$\left[\tfrac{1}{2}\dot{r}^2(m,t)\right]^{\cdot} = Gm[1/r(m,t)]^{\cdot},$$

or, since m and t are independent variables,

$$\left[\tfrac{1}{2}\dot{r}^2(m,t) - Gm/r(m,t)\right]^{\cdot} = 0.$$

Integrating, we have

$$\tfrac{1}{2}\dot{r}^2(m,t) - Gm/r(m,t) = -C,$$

where $-C$ is an integration constant (independent of time). If the collapse starts from rest, that is, $\dot{r}(m,0) = 0$ everywhere, then $C = Gm/r(m,0)$. We choose $C = 0$, implying collapse from a very extended initial configuration. All solutions will converge with time to that corresponding to $C=0$, since the term $Gm/r(m,t)$, which increases with time, will eventually become dominant. For a uniform density, $m = \frac{4\pi}{3} r(m,t)^3 \rho$, where $\rho = \rho(t)$, and hence

$$\dot{r}^2(m,t) = \frac{8\pi G \rho(t)}{3} r^2(m,t). \qquad \text{(Ex. 9.1.1)}$$

Therefore

$$\dot{r}(m,t) = -\sqrt{8\pi G \rho(t)/3}\, r(m,t),$$

where we have chosen the negative root, appropriate to collapse. This shows that at any given time the velocity changes linearly with distance from the centre.

Note: The same equation of motion applies to the universe (in the Newtonian approach) and describes its expansion – when the positive root of equation (Ex. 9.1.1) is chosen. The resulting linear dependence of velocity on distance – describing the relative motion of galaxies – is known as the *Hubble law*, which was first discovered from observations.

Exercise 9.2: We proceed as in Exercise 9.1 to obtain the first integral of the equation of motion,

$$\tfrac{1}{2}\dot{r}^2 = Gm\left(\frac{1}{r} - \frac{1}{r_0}\right),$$

where $r \equiv r(m,t)$ and $r_0 \equiv r(m,0)$. From the condition of uniform initial density, which we denote by ρ_0, we have

$$m = \frac{4\pi}{3} r_0^3 \rho_0.$$

Substituting m in the former relation, we obtain

$$\dot{r} = -\left[\frac{8\pi G \rho_0 r_0^2}{3}\left(\frac{r_0}{r} - 1\right)\right]^{1/2}, \qquad \text{(Ex. 9.2.1)}$$

where we have chosen the negative root to describe the collapse. In order to solve this equation, we introduce a new variable, $x(m,t)$, defined by

$$\cos^2 x = \frac{r}{r_0},$$

noting that $x = 0$ at $t = 0$. We also define a constant $K = \sqrt{8\pi G \rho_0/3}$. It is easy to see that equation (Ex. 9.2.1) becomes

$$\dot{x} \cos^2 x = \tfrac{1}{2} K,$$

which may be directly integrated to yield

$$x + \tfrac{1}{2}\sin 2x = Kt.$$

Now, the solution $x(t)$, or r/r_0, is the same for all m, meaning that any part of the core will take the same amount of time to contract to a given fraction of its former radial distance from the centre. The density will thus remain uniform. It is noteworthy that the time of collapse is finite: when $r(m,t) = 0$, $x = \pi/2$ and $t = 2K/\pi$ (which is of the order of the dynamical timescale $1/\sqrt{G\rho_0}$). Hence the solution has a singularity, the density becoming infinite at $t = 2K/\pi$.

Exercise 10.1: Consider a cloud of mass equal to the Jeans mass M_J and temperature T. According to equation (10.4), its radius is

$$R = \frac{\alpha}{3} \frac{\mu G M}{\mathcal{R} T}. \qquad \text{(Ex. 10.1.1)}$$

The rate of gravitational energy release in collapse may be estimated by the potential gravitational energy, of the order of GM_J^2/R [equation (2.27)], divided by the free-fall, or dynamical timescale, [equation (2.56)]. Thus,

$$\dot{E}_{\text{grav}} \approx \alpha \frac{GM_J^2}{R} \left(\frac{2GM_J}{R^3} \right)^{1/2} \approx \alpha \sqrt{2} G^{3/2} M_J^{5/2} R^{-5/2}.$$

Since the radiation temperature is lower than the gas temperature T, the rate at which energy is radiated at the cloud's surface, or the cloud's luminosity L, may be taken as

$$L = \epsilon\, 4\pi R^2 \sigma T^4,$$

where $\epsilon < 1$. As the radiated energy is supplied by the gravitational energy released in collapse, we have

$$\epsilon\, 4\pi R^2 \sigma T^4 = \alpha \sqrt{2} G^{3/2} M_J^{5/2} R^{-5/2}. \qquad \text{(Ex. 10.1.2)}$$

Substituting equation (Ex. 10.1.1) into equation (Ex. 10.1.2) yields

$$M_J \approx \left[\frac{4.0(\mathcal{R}/\mu)^{9/4}}{\sigma^{1/2} G^{3/2}} \right] \frac{T^{1/4}}{\epsilon^{1/2}}.$$

Since $\epsilon < 1$, taking $\epsilon = 1$ on the right-hand side provides a lower limit for M_J.

Exercise 10.2: From equation (10.13) we have

$$\frac{\eta}{\zeta} = \frac{\left(M_{\text{MS}}^{-0.35} - M_{\text{max}}^{-0.35}\right) - M_{\text{WD}}\left(M_{\text{MS}}^{-1.35} - M_{\text{SN}}^{-1.35}\right)\frac{0.35}{1.35}}{M_{\text{min}}^{-0.35} - M_{\text{max}}^{-0.35}}.$$

Substituting $M_{MS} = 0.7 M_\odot$, $M_{SN} = 10 M_\odot$, and $M_{WD} = 0.6 M_\odot$, we obtain

$$\frac{\eta}{\zeta} = \frac{0.89(M_{max}/M_\odot)^{0.35} - 1}{(M_{max}/M_{min})^{0.35} - 1},$$

which yields for $M_{max} = 30 M_\odot$

$$\frac{\eta}{\zeta} = 0.23 \quad (M_{min} = 0.05 M_\odot) \qquad \frac{\eta}{\zeta} = 0.40 \quad (M_{min} = 0.20 M_\odot),$$

while for $M_{max} = 120 M_\odot$

$$\frac{\eta}{\zeta} = 0.26 \quad (M_{min} = 0.05 M_\odot) \qquad \frac{\eta}{\zeta} = 0.45 \quad (M_{min} = 0.20 M_\odot).$$

In conclusion, the ratio η/ζ is far more sensitive to M_{min} than to M_{max}.

Exercise 10.3: As we have seen in Section 1.4, and again in Section 7.4, the luminosity of main-sequence stars is a function of the stellar mass in the form of a power law, $L \propto M^\nu$. If the cluster's luminosity L_C is the sum of the luminosities of its main-sequence stars, which have masses in the range $M_{min} \leq M \leq M_{tp}$, then

$$L_C = \int_{M_{min}}^{M_{tp}} L(M) dN = \int_{M_{min}}^{M_{tp}} L(M)\Phi(M) dM \propto \int_{M_{min}}^{M_{tp}} M^{\nu - 2.35} dM.$$

The relative change in L_C from $L_{C,1}$, say, to $L_{C,2}$, as the turning point off the main sequence decreases from $M_{tp,1} = 1.3 M_\odot$ to $M_{tp,2} = 0.85 M_\odot$, is given by

$$\frac{L_{C,2}}{L_{C,1}} = \frac{\int_{M_{min}}^{M_{tp,2}} M^{\nu - 2.35} dM}{\int_{M_{min}}^{M_{tp,1}} M^{\nu - 2.35} dM} = \frac{(M_{tp,2}/M_{min})^{\nu - 1.35} - 1}{(M_{tp,1}/M_{min})^{\nu - 1.35} - 1} \approx \left(\frac{M_{tp,2}}{M_{tp,1}}\right)^{\nu - 1.35}.$$

Thus the cluster's luminosity decreases by a factor of ~ 2, if we adopt $\nu = 3$, and by a factor of ~ 5, if $\nu = 5$.

Exercise 10.4: The function $\Upsilon(t)$ satisfies the equation

$$\dot{\Upsilon} = -\alpha \Upsilon^2,$$

where α is a constant to be determined from the given data. Integrating and using the initial condition $\Upsilon(0) = 1$, we have

$$\frac{1}{\Upsilon} - 1 = \alpha t.$$

Substituting $\Upsilon(t_p) = \Upsilon_p = 0.05$, we may eliminate α to obtain

$$\Upsilon(t) = \frac{\Upsilon_p}{\Upsilon_p + \frac{t}{t_p}(1 - \Upsilon_p)}.$$

Substituting $\Upsilon = 0.5$ yields $t/t_p = 0.053$, meaning that when the Galaxy was ~5% of its present age, the gas content amounted to half the galactic mass. It decreased to a tenth of the galactic mass when the Galaxy reached about half its present age. Decreasing further, it will reach half its present mass (that is, $\Upsilon = 0.025$) when the Galaxy will be about twice its present age.

Appendix 3

Physical and astronomical constants

Table A3.1 Fundamental constants

Constant	Symbol	Value	SI	cgs
			Units	
Speed of light	c	2.99792458	10^8 m s^{-1}	10^{10} cm s^{-1}
Permittivity	ε_0	$1/4\pi$	$10^7/c^2$ C^2 N^{-1} m^{-2}	1
Permeability	μ_0	4π	10^{-7} C^{-2}N s^2	1
Gravitational	G	6.67259	10^{-11} m^3 kg^{-1} s^{-2}	10^{-8} cm^3 g^{-1} s^{-2}
Planck	h	6.6260755	10^{-34} J s	10^{-27} erg s
Boltzmann	k	1.380658	10^{-23} J K^{-1}	10^{-16} erg K^{-1}
Stefan–Boltzmann	σ	5.67051	10^{-8} J m^{-2} s^{-1} K^{-4}	10^{-5} erg cm^{-2} s^{-1} K^{-4}
Radiation	a	7.5646	10^{-16} J m^{-3} K^{-4}	10^{-15} erg cm^{-3} K^{-4}
Wien		2.897756	10^{-3} m K	10^{-1} cm K
Avogadro	N_A	6.0221367	10^{23} mol^{-1}	10^{23} mol^{-1}
Atomic mass unit	m_H	1.6605402	10^{-27} kg	10^{-24} g
Ideal gas	$\mathcal{R}$	8.314510	10^3 J kg^{-1} K^{-1}	10^7 erg g^{-1} K^{-1}
Electron charge	e	1.60217733	10^{-19} C	$10^{-20}c$ esu
Electron mass	m_e	9.1093897	10^{-31} kg	10^{-28} g
Proton mass	m_p	1.6726231	10^{-27} kg	10^{-24} g
Neutron mass	m_n	1.6749286	10^{-27} kg	10^{-24} g
Angström	Å	1	10^{-10} m	10^{-8} cm

Note: $a = 4\sigma/c$, $m_H = 1/N_A$, $\mathcal{R} = k/m_H$. Fundamental constants are from E. R. Cohen & B. N. Taylor, (1987), *Rev. Mod. Phys.*, 59, p. 1121; *CODATA Bulletin* (1986), 63, Nov.; *Physics Today* (1995), Part 2, BG9, Aug.

Table A3.2 Astronomical constants

Constant	Symbol	Value	SI	cgs
			Units	
Solar mass	$M_\odot$	1.9891	10^{30} kg	10^{33} g
Solar radius	$R_\odot$	6.9598	10^8 m	10^{10} cm
Solar luminosity	$L_\odot$	3.8515	10^{26} J s^{-1}	10^{33} erg s^{-1}
Year (solar)	yr	3.1558	10^7 s	10^7 s
Light-year	ly	9.463	10^{15} m	10^{17} cm
Parsec	pc	3.086	10^{16} m	10^{18} cm
Astronomical Unit	AU	1.496	10^{11} m	10^{13} cm
Earth mass	$M_\oplus$	5.976	10^{24} kg	10^{27} g
Earth radius	$R_\oplus$	6.378	10^6 m	10^8 cm

Note: Astronomical constants are from C. Caso et al. (1998), *European Physical Journal*, C3, p. 1.

Table A3.3 Energy conversion factors

Units	erg	eV	s^{-1}	cm^{-1}	K
erg	1	1.60217733(−12)	6.6260755(−27)	1.9864475(−16)	1.380658(−16)
eV	6.2415064(11)	1	4.1356692(−15)	1.23984244(−4)	8.617385(−5)
s^{-1}	1.50918897(26)	2.41798836(14)	1	2.99792458(10)	2.083674(10)
cm^{-1}	5.0341125(15)	8.0655410(3)	3.335640952(−11)	1	6.950387(−1)
K	7.242924(15)	1.160445(4)	4.799216(−11)	1.438769	1

Note: Powers of 10 are given in parentheses. The units of energy are related as follows: 1 J = 10^7 erg; 1 erg = $1/e$ eV = $1/h$ s^{-1} = $1/(hc)$ cm^{-1} = $1/k$ K. Energy conversion factors are from E. R. Cohen & B. N. Taylor (1987), *Rev. Mod. Phys.*, 59, p. 1121; *CODATA Bulletin* (1986), 63, Nov.; *Physics Today*, (1995), Part 2, BG9, Aug. Values within the same column are equivalent.

Table A3.4 Mass excesses

Z	Element	A	ΔM (MeV)
0	n	1	8.071
1	H	1	7.289
	D	2	13.136
2	He	3	14.931
		4	2.425
3	Li	6	14.086
		7	14.908
4	Be	9	11.348
5	B	10	12.051
		11	8.668
6	C	12	0
		13	3.125
7	N	14	2.863
		15	0.101
8	O	16	−4.737
		17	−0.809
		18	−0.782
9	F	19	−1.487
10	Ne	20	−7.042
		21	−5.732
		22	−8.024
11	Na	23	−9.530
12	Mg	24	−13.933
		25	−13.193
		26	−16.215
13	Al	27	−17.197
14	Si	28	−21.493
		29	−21.895
		30	−24.433
15	P	31	−24.441
16	S	32	−26.016
		33	−26.586
		34	−29.932
26	Fe	56	−60.601
27	Co	59	−62.224
28	Ni	58	−60.223
		60	−64.468

Note: Published by J. K. Tuli, National Nuclear Data Center, Brookhaven National Laboratory.

Bibliography

Articles

Arnett, W. D., Bahcall, J. N., Kirshner, R. P., Woosley, S. E. (1989), *Ann. Rev. Astron. Astrophys.*, 27, pp. 629–700.
Atkinson, R. d'E., Houtermans, F. G. (1929), *Zeit. f. Physik*, 54, pp. 656–665.
Audouze, J., Tinsley, B. M. (1976), *Ann. Rev. Astron. Astrophys.*, 14, pp. 43–79.
Baade, W. (1944), *Astrophys. J.*, 100, pp. 137–150.
Baade, W., Zwicky, F. (1934), *Phys. Rev.*, 45, p. 138.
Bahcall, J. N. (1964), *Phys. Rev. Letters*, 12, pp. 300–302.
Barbon, R., Ciatti, F., Rosino, L. (1973), *Astron. Astrophys.*, 25, pp. 241–248.
Bergeron, P., Saffer, R. A., Liebert, J. (1992), *Astrophys. J.*, 394, pp. 228–247.
Bertout, C. (1989), *Ann. Rev. Astron. Astrophys.*, 27, pp. 351–395.
Bethe, H. A. (1939), *Phys. Rev.*, 55, pp. 434–456.
Bethe, H. A., Critchfield, C. H. (1938), *Phys. Rev.*, 54, pp. 248–254.
Biermann, L. (1932), *Zeits. Astrophys.*, 5, pp. 117–139.
Biermann, L. (1935), *Astron. Nachr.*, 257, pp. 269–294.
Burbidge, E. M., Burbidge, G. R., Fowler, W. A., Hoyle, F. (1957), *Rev. Mod. Phys.*, 29, pp. 547–650.
Burrows, A., Marley, M., Hubbard, W. B., Lunine, J. I., Guillot, T., Saumon, D., Freedman, R., Sudarsky, D., Sharp, C. (1997), *Astrophys. J.*, 491, pp. 856–875.
Chandrasekhar, S. (1931), *Astrophys. J.*, 74, pp. 81–82.
Chiosi, C., Bertelli, G., Bressan, A. (1992), *Ann. Rev. Astron. Astrophys.*, 30, pp. 235–285.
Chiosi, C., Maeder, A. (1986), *Ann. Rev. Astron. Astrophys.*, 24, pp. 329–375.
Colgate, S. A., White, R. H. (1966), *Astrophys. J.*, 143, pp. 626–681.
Cook, C. W., Fowler, W. A., Lauritsen, C. C., Lauritsen, T. (1957), *Phys. Rev.*, 107, pp. 508–515.
Cowling, T. G. (1930), *Mon. Not. Roy. Astron. Soc.*, 91, pp. 92–108.
Cowling, T. G. (1934), *Mon. Not. Roy. Astron. Soc.*, 94, pp. 768–782.
Cowling, T. G. (1935), *Mon. Not. Roy. Astron. Soc.*, 96, pp. 15–20.
Cowling, T. G. (1935), *Mon. Not. Roy. Astron. Soc.*, 96, pp. 42–60.
Cowling, T. G. (1966), *Quart. J. Roy. Astron. Soc.*, 7, pp. 121–137.
D'Antona, F., Mazzitelli, I. (1986), *Astron. Astrophys.*, 162, pp. 80–86.
Davis, R. Jr. (1964), *Phys. Rev. Lett.*, 12, pp. 302–305.
Doggett, J. B., Branch, D. (1985), *Astron. J.*, 90, pp. 2303–2311.

Dunbar, D. N. F., Pixley, R. E., Wenzel, W. A., Whaling, W. (1953), *Phys. Rev.*, 92, pp. 649–650.
Eddington, A. S. (1916), *Mon. Not. Roy. Astron. Soc.*, 77, pp. 16–35.
Eddington, A. S. (1932), *Mon. Not. Roy. Astron. Soc.*, 92, pp. 471–481.
Eggleton, P. P., Cannon, R. C. (1991), *Astrophys. J.*, 383, pp. 757–760.
Eggleton, P. P., Faulkner, J., Cannon, R. C. (1998), *Mon. Not. Roy. Astron. Soc.*, 298, pp. 831–834.
Faulkner, J. (1966), *Astrophys. J.*, 144, pp. 978–994.
Fowler, R. H., Guggenheim, E. A. (1925), *Mon. Not. Roy. Astron. Soc.*, 85, pp. 939–960, 961–970.
Gamow, G. (1928), *Zeit. f. Physik*, 52, pp. 510–515.
Gamow, G., Teller, E. (1938), *Phys. Rev.*, 53, pp. 608–609.
Gautschy, A., Saio, H. (1996), *Ann. Rev. Astron. Astrophys.*, 34, pp. 551–606.
Gold, T. (1968), *Nature*, 218, pp. 731–732.
Gold, T. (1969), *Nature*, 221, pp. 25–27.
Hamada, T., Salpeter, E. E. (1961), *Astrophys. J.*, 134, pp. 683–698.
Härm, R., Schwarzschild, M. (1966), *Astrophys. J.*, 145, pp. 496–504.
Haselgrove, C. B., Hoyle, F. (1956), *Mon. Not. Roy. Astron. Soc.*, 116. pp. 515–526.
Haxton, W. C. (1995), *Ann. Rev. Astron. Astrophys.*, 33, pp. 459–503.
Hayashi, C. (1966), *Ann. Rev. Astron. Astrophys.*, 4, pp. 171–192.
Hayashi, C., Hoshi, R., Sugimoto, D. (1962), *Progr. Theor. Phys. Suppl.*, 22, pp. 1–183.
Henyey, L. G., Forbes, J. E., Gould, N. L. (1964), *Astrophys. J.*, 139, pp. 306–317.
Hewish, A., Bell, S. J., Pilkington, J. D. H., Scott, P. F., Collins, R. A. (1968), *Nature*, 217, pp. 709–713.
Hoyle, F. (1946), *Mon. Not. Roy. Astron. Soc.*, 106, pp. 343–383.
Hoyle, F. (1953), *Astrophys. J.*, 118, pp. 513–528.
Hoyle, F. (1954), *Astrophys. J. Suppl.*, 1, pp. 121–146.
Hoyle, F. (1960), *Mon. Not. Roy. Astron. Soc.*, 120, pp. 22–32.
Hoyle, F., Lyttleton, R. A. (1939), *Proc. Camb. Phil. Soc.*, 35, pp. 592–609.
Hoyle, F., Lyttleton, R. A. (1948), *Occ. Notes Roy. Astron. Soc.*, 12, pp. 89–108.
Hoyle, F., Schwarzschild, M. (1955), *Astrophys. J. Suppl.*, 2, pp. 1–40.
Iben, I. (1965), *Astrophys. J.*, 141, pp. 993–1018.
Iben, I. (1967), *Ann. Rev. Astron. Astrophys.*, 5, pp. 571–626.
Iben, I. (1974), *Ann. Rev. Astron. Astrophys.*, 12, pp. 215–256.
Iben, I. (1985), *Quart. J. Roy. Astron. Soc.*, 26, pp. 1–39.
Iben, I. (1991), *Astrophys. J. Suppl.*, 76, pp. 55–114.
Iglesias, C. A., Rogers, J. (1996), *Astrophys. J.*, 464, pp. 943–953.
Jeans, J. H. (1902), *Phil. Trans. Roy. Soc.*, 199, pp. 1–5.
Johnson, H. L. (1952), *Astrophys. J.*, 116, pp. 272–282.
Johnson, H. L., Morgan, W. W. (1953), *Astrophys. J.*, 117, pp. 313–352.
Johnson, H. L., Sandage, A. R. (1956), *Astrophys. J.*, 124, pp. 379–389.
Kaniel, S., Kovetz, A. (1967), *Phys. Fluids*, 10, pp. 1186–1193.
Kippenhahn, R. Thomas, H.-C., Weigert, A. (1965), *Zeits. Astrophys.*, 61, pp. 241–267.
Koester, D. (1987), *Astrophys. J.*, 322, pp. 852–855.
Kovetz, A. (1969), *Mon. Not. Roy. Astron. Soc.*, 144, pp. 459–460.
Kovetz, A. (1969), *Astrophys. Sp. Sci.*, 4, pp. 365–369.
Kovetz, A., Shaviv, G. (1970), *Astron. Astrophys.*, 8, pp. 398–403.
Kovetz, A., Shaviv, G. (1976), *Astron. Astrophys.*, 52, pp. 403–407.
Kramers, H. A. (1923), *Phil. Mag.*, 46, pp. 836–871.
Kroupa, P., Tout, C. A., Gilmore, G. (1990), *Mon. Not. Roy. Astron. Soc.*, 244, pp. 76–85.

Kudritzki, R. P., Reimers, D. (1978), *Astron. Astrophys.*, 70, pp. 227–239.
Landau, L. D. (1932), *Soviet Physics*, 1, pp. 285–287.
Landau, L. D. (1938), *Nature*, 141, pp. 333–334.
Lee, T. D. (1950), *Astrophys. J.*, 111, pp. 625–640.
Low, C., Lynden-Bell, D. (1976), *Mon. Not. Roy. Astron. Soc.*, 176, pp. 367–390.
Maeder, A., Conti, P. S., (1994), *Ann. Rev. Astron. Astrophys.*, 32, pp. 227–275.
McCray, R. (1993), *Ann. Rev. Astron. Astrophys.*, 31, pp. 175–216.
McCrea, W. H. (1939), *Occ. Notes Roy. Astron. Soc.*, 1, pp. 78–88.
McCrea, W. H. (1957), *Mon. Not. Roy. Astron. Soc.*, 117, pp. 562–578.
Mestel, L. (1952), *Mon. Not. Roy. Astron. Soc.*, 112, pp. 583–597.
Mestel, L. (1952), *Mon. Not. Roy. Astron. Soc.*, 112, pp. 598–605.
Mestel, L. (1965), in *Stars and Stellar Systems*, Vol. VIII, pp. 297–325.
Mestel, L. (1965), *Quart. J. Roy. Astron. Soc.*, 6, pp. 161–198.
Mestel, L., Ruderman, M. A. (1967), *Mon. Not. Roy. Astron. Soc.*, 136, pp. 27–38.
Meynet, G., Maeder, A. (1992), *Astron. Astrophys. Suppl. Ser.*, 96, pp. 269–331.
O'Dell, C. R. (1963), *Astrophys. J.*, 138, pp. 67–78.
Oppenheimer, J. R., Volkoff, G. M. (1939), *Phys. Rev.*, 55, pp. 374–381.
Pacini, F. (1967), *Nature*, 216, pp. 567–568.
Paczyński, B. (1971), *Acta Astron.*, 21, pp. 271–288.
Parker, D., Bahcall, J. N., Fowler, W. A. (1964), *Astrophys. J.*, 139, pp. 602–621.
Prialnik, D., Shaviv, G. (1980), *Astron. Astrophys.*, 88, pp. 127–134.
Rakavy, G. Shaviv, G., (1968), *Astrophys. Sp. Sci.*, 1, pp. 429–441.
Rakavy, G., Shaviv, G., Zinamon, Z. (1967), *Astrophys. J.*, 150, pp. 131–162.
Rana, N. C. (1987), *Astron. Astrophys.*, 184, pp. 104–118.
Reimers, D. (1975), *Mem. Soc. Roy. Sci. Liège*, 8, pp. 369–382.
Renzini, A., Fusi Pecci, F. (1988), *Ann. Rev. Astron. Astrophys.*, 26, pp. 199–244.
Rosseland, S. (1924), *Mon. Not. Roy. Astron. Soc.*, 84, pp. 525–528.
Salpeter, E. E. (1952), *Astrophys. J.*, 115, pp. 326–432.
Salpeter, E. E. (1955), *Astrophys. J.*, 121, pp. 161–167.
Salpeter, E. E. (1957), *Phys. Rev.*, 107, pp. 516–525.
Salpeter, E. E. (1961), *Astrophys. J.*, 134, pp. 669–682.
Salpeter, E. E. (1966), in *Perspectives in Modern Physics, Essays in Honor of Hans A. Bethe*, ed. R. E. Marshak, Interscience, New York, pp. 463–475.
Salpeter, E. E. (1974), *Astrophys. J.*, 193, pp. 585–592.
Sandage, A. R., Schwarzschild, M. (1952), *Astrophys. J.*, 116, pp. 463–476.
Sandage, A., Tammann, G. A. (1968), *Astrophys. J.*, 151, pp. 531–545.
Schönberg, M., Chandrasekhar, S. (1942), *Astrophys. J.*, 96, pp. 161–172.
Schwarzschild, K. (1906), *Göttinger Nachr.*, 195, pp. 41–53.
Schwarzschild, M., Härm, R. (1965), *Astrophys. J.*, 142, pp. 855–867.
Shaviv, G., Kovetz, A. (1972), *Astron. Astrophys.*, 16, pp. 72–76.
Shaviv, G., Kovetz, A. (1976), *Astron. Astrophys.*, 51, pp. 383–391.
Stevenson, D. (1991), *Ann. Rev. Astron. Astrophys.*, 29, pp. 163–193.
Strömgren, B. (1932), *Zeit. Astrophys.*, 4, pp. 118–152.
Strömgren, B. (1933), *Zeit. Astrophys.*, 7, pp. 222–248.
Sweeney, M. A. (1976), *Astron. Astrophys.*, 49, pp. 375–385.
Tayler, R. J. (1952), *Mon. Not. Roy. Astron. Soc.*, 112, pp. 387–398.
Tayler, R. J. (1954), *Astrophys. J.*, 120, pp. 332–341.
Tayler, R. J. (1956), *Mon. Not. Roy. Astron. Soc.*, 116, pp. 25–37.
Weaver, T. A., Woosley, S. E. (1980), *Ann. NY Acad. Sci.*, 336, pp. 335–357.
Weidemann, V. (1990), *Ann. Rev. Astron. Astrophys.*, 28, pp. 103–137.

Weidemann, V., Koester, D. (1984), *Astron. Astrophys.*, 132, pp. 195–202.
Weizsäcker, C. F. von (1937), *Physik. Zeit.*, 38, pp. 176–191.
Weizsäcker, C. F. von (1938), *Physik. Zeit.*, 39, pp. 633–646.
Wheeler, J. A. (1966), *Ann. Rev. Astron. Astrophys.*, 4, pp. 393–432.
Winget, D. E., Hansen, C. J., Liebert, J., Van Horn, H. M., Fontaine, G., Nather, R. E. Kepler, S. O., Lamb, D. Q. (1987), *Astrophys. J.*, 315, pp. L77–L81.
Woosley, S. E., Weaver, T. A. (1986), *Ann. Rev. Astron. Astrophys.*, 24, pp. 205–253.
Zapolski, H. S., Salpeter, E. E. (1969), *Astrophys. J.*, 158, pp. 809–813.

Books

Arnett, D. (1996), *Supernovae and Nucleosynthesis: An Investigation of the History of Matter, from the Big Bang to the Present*, Princeton University Press, Princeton.
Bahcall, J. N. (1989), *Neutrino Astrophysics*, Cambridge University Press, Cambridge.
Barnes, C. A., Clayton, D. D. and Schramm, D. D., eds. (1982), *Essays in Nuclear Astrophysics*, Cambridge University Press, New York.
Chandrasekhar, S. (1939), *An Introduction to the Study of Stellar Structure*, Dover, New York.
Clayton, D. D. (1968), *Principles of Stellar Evolution and Nucleosynthesis*, McGraw-Hill, New York.
Eddington, A. S. (1926), *The Internal Constitution of the Stars*, Cambridge University Press, Cambridge.
Kippenhahn, R. (1983), *100 Billion Suns: The Birth, Life and Death of the Stars*, Princeton University Press, Princeton.
Kippenhahn, R. and Weigert, A. (1990), *Stellar Structure and Evolution*, Springer-Verlag, Berlin.
Menzel, D. H., Bhatnagar, P. L. and Sen, H. K. (1963), *Stellar Interiors*, Wiley, New York.
Novotny, E. (1973), *Introduction to Stellar Atmospheres and Interiors*, Oxford University Press, New York.
Rybicki, G. B. and Lightman, A. P. (1979), *Radiation Processes in Astrophysics*, Wiley, New York.
Schatzman, E. L. (1958), *White Dwarfs*, Interscience, New York.
Schatzman, E. L. and Praderie, F. (1993), *The Stars*, Springer-Verlag, Berlin.
Schwarzschild, M. (1958), *Structure and Evolution of the Stars*, Princeton University Press, Princeton.
Shapiro, S. L. and Teukolsky, S. A. (1983), *Black Holes, White Dwarfs and Neutron Stars*, Wiley, New York.
Stein, R. F. and Cameron, A. G. W., eds. (1966), *Stellar Evolution*, Plenum, New York.
Tayler, R. J. (1994), *The Stars: Their Structure and Evolution*, Cambridge University Press, Cambridge (first published 1970).
Tayler, R. J. (1997), *The Sun as a Star*, Cambridge University Press, Cambridge.
Tinney, C. G., ed. (1994), *The Bottom of the Main Sequence – and Beyond*, Springer-Verlag, Berlin.

Index

Note: Reference to the subject of a Section is given in italics. Page numbers followed by f or t refer to figures or tables, respectively.

absorption, 4, 44, 48, 49, 50, 51, 216, 217, 218f
accretion, 180, *192–4*, 198, 200
   rate, 193
adiabatic exponent, *45–8*, 96–7, 98, 100–4, 134, 191, 233–5
adiabatic gradient, 100, 104, 132, 235
adiabatic process, 45, 96, 99–100, 233
age
   stellar cluster, 122, 141–2, 141f
   Sun, 31, 32–3, 141
   universe, 33, 142, 193, 208
Andromeda galaxy, 176, 178t
apparent brightness, 2, 9, 142, 153, 155
Archimedes buoyancy principle (law), 100, 103
Aristotle, 177–8
Astronomical Unit (AU, Earth–Sun distance), 3, 249t
asymptotic giant branch (AGB), *155–60*, 160–2, 164
Atkinson, Robert, 58
atomic nucleus, 26–8, 55, 62, 68–9, 181

Baade, Walter, 187, 211
Bahcall, John N., 144
baryon number, 26, 55
Bell, Jocelyn, 188
β decay, 55, 62, 184
Bethe, Hans A., 59, 62, 128
Biermann, Ludwig, 103
binary star (system), 4, 6, 117, 180
black dwarf, 170, 200
black hole, 188, *191–2*, 193–4, 203, 208
blackbody radiation, 4, 16, 51, 126, 194, 215
Boltzmann formula, 217
Boltzmann constant, 22, 249t

boundary conditions, 29, 73, 75, 226–7
Brahe, Tycho, 178
bremsstrahlung, 49, 55
brown dwarf, *199–203*, 201f, 204
Burbidge, Geoffrey R., 69
Burbidge, Margaret E., 69

carbon, 6, 55, 62, 64–5, 138, 168, 171, 173, 227
   burning, *65–7*, 66f, 68t, 171
   detonation, 117, 118
carbon-oxygen core, 127, 155–8, 161, 162, 164
Cepheid, 153–4, 162, 176
   period-luminosity relation, 153–4, 153f
Chamberlin, Thomas C., 32
Chandrasekhar, Subrahmanyan, 73, 80, 87, 146, 235
   limiting (critical) mass, *79–80*, 97, 114, 117, 129, 130, 161–2, 178, 180, 187, 231, 232
charge (atomic number), 26, 36, 40, 50, 58, 68–9
Cherenkov light, 145
   water-Cherenkov experiment, 145
Clayton, Donald D., 59
cluster, stellar, 2, 9, 11–13, 11f, 122, 129, 141–2, 174, 175f, 198, 247
   globular, 150, 211
   Hyades, 12f
   M3, 12f
   Pleiades, 11f, 12f
   47 Tucanae, 11f
CNO cycle (bi-cycle), *62*, 63f, 115, 121, 139, 148, 155, 171, 236, 243
Colgate, Stirling A., 182
collapse, 31, 97, 105, 117, 135, 178, 181, 188, 189, 192, 198, 244–5
comet, 1, 178, 211

composition (abundances), 70, 73–4, 171, 173, 173f, 208
  change, equation of, *26–8*,
  cosmic, 64, 184
  initial, *6–7*, 34, 74
  mass fraction, 7, 16, 27, 28, 39, 40, 136
  solar, 7, 39, 40, 50, 184, 227
conservation laws
  angular momentum, 17, 19, 189
  baryon number, 26, 27, 54, 59
  charge, 26, 27, 59
  energy, 17, 18, 25, 31, 70, 126, 215–16
  lepton number, 26–7, 28, 59, 182
  mass, 8, 17, 18, 240
  momentum, 17, 19, 70
continuity equation, 8, 17, 72, 86, 96, 119
convection, 73, 81, *98–101*, 102–3, 132, 138, 140f, 150, 156, 175
  stability criterion, 99–101, 99f, 100f
Coulomb
  barrier, 57f, 57–8, 64, 65, 67, 68
  energy, 36, 42, 226
  field (force), 26, 55, 56
  interaction, 36, 169
Cowling, Thomas G., 86, 88, 102
  point-source model, *86–8*

dark matter (missing mass), 203, 204
Darwin, Charles, 32–3
Davis, Raymond, 144
  chlorine experiment, 144, 145
Debye temperature, 169
degeneracy pressure, 41–3, 79, 97, 102, 150, 201, 227
density
  average, 23, 30, 36, 77, 149f
  central, 77, 78, 93, 106, *111–14*, 113f, 115f, 129f, 149f, 231
deuterium, 59, 61, 201
  burning, 201
diffusion approximation, 221
dimensional analysis, 119
Doppler effect (shift), 5, 7, 203
dredge-up, 139, 150, 156, 158

Earth, 3, 10, 16, 143, 177, 183, 189, 202f, 240
Eddington, Sir Arthur S., 15, 32–3, 35–6, 51, 73, 82, 84, 85–6, 175, 220, 235
  luminosity, *81–2*, 83, 85, 158, 159, 171, 193, 235
  quartic equation, 84, 84f
  standard model, *82–5*, 88, 102, 121, 129
effective temperature, 4, 10, 122, 126, 132, 135, 136, 152, 156, 162, 169, 236

Einstein, Albert, 217
  General Theory of Relativity, 188, 192
  mass-energy relation, 33, 53
  relations (radiation coefficients), 218
electron-positron pair, 55, 70
emission, 4, 44, 49, 216, 217, 218f
  induced (stimulated), 217
  spontaneous, 217
energy
  gravitational potential (binding), 7, 21–3, 25–6, 31, 32, 90–1, 126, 135, 165, 181, 182, 190, 192–3, 198, 233, 242
  kinetic, 25, 182, 190
  nuclear, 18, 122, 181
  nuclear binding, 33, *53–5*, 56f
  radiation, 45, 91, 97, 181
  rest-mass, 30, 33, 53, 70, 188
  specific (internal, per unit mass), 18, 22, 44–5, 47, 103, 226
  thermal (internal), 17, 21, 25, 31, 32, *44–5*, 57, 90–1, 92, 126, 165, 168, 198
  total (stellar), *24–6*, 31, 32, 90, 97, 126, 148
energy equation, *17–19*, 24, 54
entropy, 46, 106
equation of motion, *19–21*, 24, 244, 245
equation of state, *36–8*, 72, 93, 94, *107–9*, 108f, 112, 167
  degenerate electron gas, 42, 79, 93–4, 107, 113, 167
  degenerate neutron gas, 187–8
  ideal gas, 40, 47, 83, 86, 93–4, 101, 107, 112, 132, 197, 227, 243
  polytropic, 75, 79, 83, 104
  radiation, 44, 83, 109, 222
  relativistic degenerate electron gas, 43, 79, 108, 231
equilibrium
  hydrostatic, 20, 25–6, 31, 34, 75, 81–2, 93, 94, 95–6, 104, 112, 135, 138, 196–7, 199
  nuclear (statistical), 28, 33, *67*, 111, 184
  radiative, 81–2, 102, 119, 139, 166, 238
  thermal, 19, 25, 32, 33, 34, 92, 94, 99, 115–16, 122, 126, 138, 148, 150, 161, 193
  thermodynamic, 15–16, 38, 43, 196, 217, 218, 219
escape velocity, 30, 159, 191, 242
evolution cycle, *208–12*, 208f
evolution equations, set of, 17, *29*, 34, 131
fermi, 26, 59
first law of thermodynamics, 17–18, 45–7
Fowler, William A., 69, 80
fragmentation, 11, 196, 198, 199f, 200f
free fall, 30, 135, 181, 244

galactic disc, 195, 203, 204, 211
galactic halo, 211
Galaxy, see Milky Way
galaxy, 2, 6, 154, 176, 203, 208, 245
γ-ray, 187, 194
Gamow, George, 58
  peak, 58
gas
  classical (nonrelativistic), 44
  degenerate electron, 42, 45, 46, 75, 79, 92, 114, 116, 156, 165–7, 180, 226, 244
  degenerate neutron, 181, 187–8
  ideal (nondegenerate), 22, 25, 36, 39, 41, 45, 46, 75, 91, 92, 114, 146, 166–7, 234
  perfect (noninteracting, free), 36, 42, 44, 169, 226
  relativistic degenerate electron, 43, 45, 46, 75, 79, 97
gas constant, 39, 249t
Gold, Thomas, 188, 190
gravitational constant, 7, 249t
gravitational force (field), 20, 31, 114, 135, 159, 192, 193, 203, 244

Härm, Richard, 95
Hayashi, Chushiro, 132
  forbidden zone, *132–6*, 150, 156
  track, 132, 134–5, 136, 139, 150, 200
heat flow (flux), 18, 52, 86, 95, 132
heavy element (metal), 6, 39, 40, 139, 152, 184, 208, 210–12, 211f
  creation, *68–70*
Heisenberg uncertainty principle, 41
helium, 4, 6, 15, 23, 39, 40, 55, 71, 102, 135, 138, 156, 171, 173, 195, 211f
  flash, 151, 152, 157
  main sequence, 152–3
helium burning, *63–5*, 68t
  core, 116, 117, 127, *151–5*, 159, 171, 191
  shell, 127, 155, 157–8, 164
Helmholtz, Hermann von, 32
Henyey, Louis, 131
  relaxation method, 131
Hertzsprung, Ejnar, 9, 154
  gap, 148, 149f
Hewish, Anthony, 188
Hipparchus of Nicea, 3
Hipparcos satellite, 3
homologous contraction, 181
homology, 120, 146
horizontal branch, 152, 155
Houtermans, Fritz, 58
Hoyle, Sir Fred, 63, 64, 69, 88, 127, 181, 188, 198

H–R diagram, *9–14*, 9f, 12f, 131–43, 137f, 139f, 141f, 148–64, 149f, 154f, 169–71, 170f, *173–5*, 174f, 175f, 183, 200, 211
Hubble, Edwin P., 176
  law, 245
hydrogen, 6, 15, 23, 39, 40, 55, 86, 102, 118, 171, 180, 199
  atomic, 23, 134, 195, 196
  molecular, 134, 135, 195
hydrogen burning, *59–63*, 68t
  core, 115, 117, 118, 138, 148, 191
  shell, 125, 127, 149, 152, 157–8, 164, 165
hydrogen 21-cm radio line, 196
hydrostatic (equilibrium) equation, 20, 72, 86, 95, 102, 112, 119, 133, 166, 224, 225, 226, 232, 243

Iben, Icko Jr., 131
infrared radiation, 4
initial mass function, *203–7*, 205f, 210
instability
  convective, *101–4*
  dynamical, *97–8*, 111, 112f, 178, 191
  Jeans, 197
  Schönberg–Chandrasekhar, 146, 150
  thermal, of degenerate gas, *92–4*, 111, 116, 150–1
  thin shell, *94–5*, 157
intensity of radiation, 215
interstellar medium, 6, 184, 186, *195–6*, 206, 208
ionization, 23, 39, 41, 46, 50, 98, 101, 102, 135–6, 152, 162, 195–6
  degree, 47
  potential, 47, 70, 135
iron, 55, 67, 71, 85, 117, 171
  core, 171, 173f, 178, 180, 182, 184, 188
  group, 67, 171
  photodisintegration, *70–1*, 111, 117, 118, 181
isothermal core, 146–8, 156, 166, 168

Jeans, Sir James H., 197
  mass, 197–8, 246
  minimum mass, 198, 199
  radius, 197
Jupiter, 36, 201f, 202, 202f

Kaniel, Shmuel, 101
Kelvin, Lord (William Thomson), 32–3
  Kelvin–Helmholtz timescale, *31–3*, 99, 104, 136, 148, 242
Kepler, Johannes, 178
  third law, 223

Kippenhahn, Rudolf, 174
Kirchhoff's law, 217, 218
Kovetz, Attay, 101, 107
Kramers, Hendrick A., 50, 85
  opacity law, 50, 102, 124, 167, 237–8

Landau, Lev, 187
Lane–Emden equation, 76, 229–30
Laplace, Pierre-Simon, Marquis de, 192
Large Magellanic Cloud (LMC), 178t, 179f, 183, 185f
Leavitt, Henrietta Swan, 153
Lee, Tsung-Dao, 93
Legendre polynomials, 219
lepton number, 26
light-year, 3, 249t
local thermodynamic equilibrium (LTE), *16–17*, 219
luminosity, 2, 10, 32, 51, 83, 121–3, 126–7, 136, 151–62, 167–70, 170f, 176, 182, 201f, 209, 236
  accretion, 193–4
  Eddington critical, *81–2*, 83, 85, 158, 159, 171, 193, 235
  neutrino, 144, 241
  nuclear, 19, 33, 92, 95, 126–7
  solar, 2, 249t
Lyttleton, Raymond A., 88

Magellanic Clouds, 155
magnetic dipole radiation, 190–1
magnetic field, 8, 189, 190–1
main sequence, 10, 11–13, 82, *118–24*, 129, 136, *138–43*, 236
  fitting, 142
  lifetime, 122, 141f, 142t
  slope, 118, 122
  turning point, 141f, 206–7, 247
  zero-age, 124
Mars, 10
mass, atomic, 27, 36, 68–9
  excess, 54, 61, 250t
  mean, 39, 168
  unit, 27, 54, 249t
mass, space variable, 8, 16, 20, 52
mass, stellar, 4, 198
  as evolutionary parameter, 106, 111–14, 115–17, 134–8, 140–3, 142t, 169
  as structural parameter, 77–8, 84, 102, 119–23, 140f, 146–8, 157, 167–8, 228, 236
  initial, 7, 10, 13, 137f, 152, 154f, 161, 184, 205, 208
  solar, 4, 249t

mass limit
  Chandrasekhar, *79–80*, 97, 114, 117, 129, 130, 161–2, 178, 180, 187, 231, 232
  Jeans, 197–8, 199, 246
  main sequence, lower, 122–3, 200
  main sequence, upper, 82, 237
  neutron star, 187–8
  Schönberg–Chandrasekhar, 147–8
mass loss, 85, 129–30, 140, 151, 152, 159, 162, 171, 175, 239, 242
  rate, 159–60, 161
mass-luminosity relation, 13, 13f, 82, 85–6, 88, 118, 121, 124, 204, 236–7, 239, 247
mass-radius relation, 78, 79, 80f, 121, 187, 202f, 237, 238
Maxwellian velocity distribution, 15, 38, 41, 56
McCrea, Sir William H., 146
mean free path, 16, 49, 51, 55, 143, 196
mean free (collision) time, 16, 63
mean molecular weight, 40, 85, 146
Mestel, Leon, 93, 129, 165, 166
Milky Way (Galaxy), 2, 129, 154, 177, 186, 190, 194, 195, 210, 248
Mira variable, 162
molecular cloud, 196

nebula, 16, 136, 195, 209
  Crab, 178, 179f, 188, 190–1
  Eagle, 200f
neon, 6, 173, 184
Neptune, 202f
neutrino, 26, 27, 54–5, 60–1, 62, 74, 130, 131, 150, 171, 182, 183, 240–1
  photoneutrino, 55
  solar, *143–6*, 144t
  solar neutrino problem, 144–6
neutrino-antineutrino pair, 55
neutrino experiments
  BOREXINO, 146
  chlorine, 144, 145
  GALLEX, 145
  IMB, 183
  Kamiokande, 144–5, 183
  SAGE, 145
  SNO, 146
neutron capture, 68–9, 157
neutron star, *187–9*, 193–4, 207, 208
Newton, Sir Isaac, 18, 53
  gravitation theory, 188, 192, 245
  second law, 19, 51
nitrogen, 6, 62, 138, 156, 171
Nobel Prize, 62, 80, 188
nova, 93, 176, 208
  T Pyxidis, 209f

nuclear burning, general, 68t, 81, 98, 102–3, 106, *109–11*, 110f, 157, 162, 171, 239
nuclear energy production (release)
    rate, 18, 54, 59, 73, 81–2, 95, 102, 110
    yield, 61, 64, 66
nuclear reaction, 26–8, 53–4, 59–67, 68–9, 88
    branching ratio, 61, 66, 143, 240–1
    rate, 28, 54, *56–9*, 92, 110
    resonant, 58, 64
    r-process, *68–70*, 69f
    s-process, *68–70*, 69f, 157
nucleosynthesis, 59, 70, 111, *184–7*

opacity coefficient, 48, *49–51*, 51f, 73, 81–2, 83, 85, 86, 99, 102, 132, 136, 165, 198, 202, 216, 220
    bound-bound, 50
    bound-free, 49
    electron-scattering, 49, 50, 82, 193, 237
    free-free, 49
    Kramers law, 50, 102, 124, 167, 237–8
    reduced, 219
    Rosseland mean, 222
Oppenheimer, Robert J., 187
optical depth, 49, 132, 227
oxygen, 6, 62, 65, 155, 168, 171, 173, 184, 191, 227
    burning, *65–7*, 66f, 68t, 191

Pacini, Franco, 191
Paczyński, Bohdan, 158
pair annihilation, 55, 70
pair production, *70*, 98, 111, 117, 191
parallax, 2, 3f
parsec, 3, 249t
Pauli exclusion principle, 41
photosphere, 4, 40, 49, 132, 135, 199
photodisintegration, 67, 98
    iron, *70–1*, 111, 117, 118, 181
photoionization, 49, 67
Planck constant, 15, 249t
Planck distribution function, 16, 44, 50, 70, 217, 219
planet, 1, 36, 136, 177, *199–203*, 201f, 204, 211
planetary nebula, *162–4*, 163f, 164f, 169, 195, 208
    Helix nebula, 163f
    Henize 1357 nebula, 209f
planetary nebula nucleus (PNN), 162, 164, 164f
plasma, 24, 55
plasmon, 55
polytrope, 75–6, 76f, 79, 104, 132, 135, 169, 187, 232

polytropic index, 75, 78, 79, 104, 132, 134–5
Population I, II, 210–11
$p - p$ chain, *59–61*, 60f, 115, 121, 123, 143, 144t, 145, 240–1
pre-main-sequence, *135–8*, 137f
pressure, 20, 31, 83
    central, 21, 78, 93, 112, 231
    electron, *39–43*
    gas, 38, 83, 84, 102
    ion, *38–9*
    radiation, 38, *43–4*, 51, 81, 83–5, *97*, 101, 130, 157, 159, 193, 221, 226
pressure integral, 37–8, 42, 43, 44
proton mass, 26, 27, 42, 249t
protostar, 135–6, 199
pulsar, *188–91*, 190f
    Crab, 188, 190–1
    Vela, 188

quasi-static process, 31, 32, 72, 126

radiation
    energy, 15, 44, 97
    field, 51, 54, 67, 159, 217
    flux, 48–9, 51–2, 221
    momentum, 15, 44, 51, 159, 221
    pressure, *see* pressure, radiation
    sources, *193–4*
    temperature, 16, 194
radiation constant, 44, 249t
radiative diffusion, 72, 81, 235
radiative transfer equation, *48–52*, 72, 102, 119, 166, *215–22*
radiative zero solution, 167
radioactive isotope, 55, 68
    aluminium, 186–7
    argon, 144
    boron, 61
    cobalt, 184
    germanium, 145
    nickel, 184
radius
    Jeans, 197
    Schwarzschild, 191, 193
    solar, 4, 249t
    stellar, 4, 8, 10, 30, 49, 76, 121, 132, 136, 193
Rakavy, Gideon, 106
recombination, 49, 98, 196
red dwarf, 142, 200
red giant, 10, 12, 118, 126–7, *146–51*, 160
    branch, 127, 148, 150, 152, 156
Reimers, Dieter, 160
    formula, 160, 161

Rosseland, Svein, 222
  mean opacity, 222
rotation, 7, 152, 189
RR Lyrae variable, 152, 162
Ruderman, Malvin, 165
Russell, Henry Norris, 9
Ryle, Martin, 188

Saha, Meghnad, 47
  equation, 47
Salpeter, Edwin E., 63, 64, 127, 204
  birth function, 204
Sandage, Allan R., 127
Saturn, 201f, 202f
Schönberg, M., 146
  Schönberg–Chandrasekhar instability, 146, 150
Schatzman, Evry, 93
Schwarzschild, Karl, 99, 191
  convective stability criterion, 99–101, 99f, 100f
  radius, 191, 193
Schwarzschild, Martin, 14, 95, 126, 127
self-gravity, 1, 188, 198
Shaviv, Giora, 106
shell flash, 158
shock wave, 183, 184, 196
silicon, 67, 173
  burning, 67, 68t
Small Magellanic Cloud (SMC), 153
spectrum, 4, 6, 10, 139, 156, 180, 186
speed of light, 3, 15, 43, 145, 191, 249t
spherical symmetry, 1, 7–8, 17, 20, 74, 95, 240
stability
  convective, 101, 102
  dynamical, 95–7, 104, 134, 196
  secular (thermal), 90–2, 94
standard candle, 153
star, 1–2, 199–203
  compact, 4, 8, 31, 118, 193–4, 209
  fully convective, 100, 132, 134, 139, 238
  intermediate-mass, 143, 149f, 152, 156f, 174f, 178, 205
  low-mass, 102, 116, 121, 132, 138, 143, 150, 152, 162, 165, 174f, 201f, 204, 208
  main-sequence, 10, 13f, 118, 138, 140f, 148, 193, 204, 205f, 206–7, 247
  massive, 85, 116–17, 121, 129–30, 132, 140, 143, 170–3, 174f, 191–2, 203, 209f, 210, 243
star formation, 180, 196–9, 203, 208, 211
stars, individual
  Capella, 79, 231

Eta Carinae, 171, 172f
Proxima Centauri, 3, 6
Stefan–Boltzmann constant, 4, 249t
stellar model, 73–4
  point-source (Cowling), 86–8
  polytropic, 74–9
  standard (Eddington), 82–5, 88, 102, 121, 129
Strömgren, Bengt, 86, 88
strong interaction, 59
strong nuclear force, 26, 56–7, 59, 67, 188
structure equations, 72–3, 74, 131, 166, 238
Sun, 2, 3, 7, 8, 10, 16, 21, 30, 32, 33, 86, 123, 139, 140, 143, 211, 222, 240–1
superadiabaticity, 103, 132, 134
supergiant, 127, 128, 156, 159
supernova, 31, 118, 130, 143, 173, 176–80, 177f, 180–4, 191, 196, 205–7
  historical, 178t
  light curve, 180, 183f, 184–6, 186f, 187f
  nucleosynthesis, 184–7
  progenitor, 128, 173, 173f, 180, 184, 185f
  remnant, 178, 179f, 188, 195
  Type I, 118, 180, 186
  Type II, 118, 180, 183
supernovae, individual
  N132D, 179f
  SN1969I, 183f
  SN1987A, 183–4, 185f, 209f
  yr 1054 (Crab), 178, 188
superwind, 160–2
synchrotron radiation, 189

T Tauri star, 136
Tayler, Roger J., 102
Taylor expansion, 228
temperature, 15–16
  average, 23, 26, 57, 92, 136
  central, 93, 106, 111–14, 113f, 115f, 123, 129f, 149f, 167–9, 227
temperature-density diagram, 107–11, 108f, 110f, 112f, 113f, 115f, 124–8, 125f, 128, 129f, 200
thermal equilibrium equation, 19, 72, 86, 119
thermal pulse, 155–8, 158f, 161, 164
thermonuclear reaction, 28, 59
thermonuclear runaway, 92–3, 95, 105, 117
threshold temperature, 58, 110–11, 115–17, 118, 125, 127, 199, 200
timescale, 30
  dynamical, 30–1, 99, 104, 135, 184, 191, 198, 246
  nuclear, 33–4, 136
  thermal (Kelvin–Helmholtz), 31–3, 99, 104, 136, 148, 242

triple-α (3α) reaction, *63–64*, 65f, 127, 171
tunnelling, 58

Uranus, 202f
UV radiation, 183, 194, 195

virial theorem, *21–3*, 25, 31, 32, 45, 91, 97, 126, 136, 146, 149, 196, 203, 244
Volkoff, George, 187

weak force, 26
weak interaction, 26, 59, 182
Weigert, Alfred, 174
Weizsäcker, Carl-Friedrich von, 62
Wheeler, John Archibald, 191
white dwarf, 10, 93, 118, 143, 151, *164–70*, 180, 193–4, 200, 206, 208, 233
   cooling, 166–9, 166f

   mass, 129–30, 162, 205
   structure, 79–80, 103, 202, 243
white dwarfs, individual
   40 Eridani B, 202f
   Sirius B, 202f
White, Richard H., 182
Wien's law, 4
   constant, 249t
wind
   solar, 16, 140, 240
   stellar, 82, 85, 130, 151, 157, 159–60, 164, 171, 208, 209f
Wolf-Rayet star, 171, 172f

X-ray, 4, 194, 222

Zeeman effect, 8
Zwicky, Fritz, 176, 187, 203